Practice Papers for SQA Exams

Higher

Physics Practice Papers

This impression was fully updated in 2017 to match the most recent versions of the SQA specifications.

001/031115

10 9 8 7 6 5 4 3 2

ISBN 9780007590940

Published by
Leckie & Leckie Ltd
An imprint of HarperCollins*Publishers*
Westerhill Road, Bishopbriggs, Glasgow, G64 2QT
T: 0844 576 8126 F: 0844 576 8131
leckieandleckie@harpercollins.co.uk www.leckieandleckie.co.uk

Special thanks to
QBS (layout and illustration); Ink Tank (cover design);
Lee Haworth Mulvey (project management); Louise Robb (copy edit); Nick Forwood (proofread)

A CIP Catalogue record for this book is available from the British Library.

Acknowledgements
Whilst every effort has been made to trace the copyright holders, in cases where this has been unsuccessful, or if any have inadvertently been overlooked, the Publishers would gladly receive any information enabling them to rectify any error or omission at the first opportunity.

Introduction

Layout of the book

This book contains three complete practice exam papers for SQA Higher Physics. The layout and question style are similar to the exam you will eventually sit. A number of the questions are set at a really challenging level. Don't let this put you off; you mustn't be worried about making mistakes or asking for help. It's much better to face the tough stuff in these practice exam papers and to find that the final SQA exam is manageable than to reach the day of the exam under-prepared.

The answer section is at the back of the book.

In the answer section you will find the correct solutions, guidance on the number of marks available for each question and some hints on how to tackle the questions. Answers to open-ended questions are not given in this book. These question types provide a great opportunity to start a discussion with your teacher over the different approaches you could take to answer open-ended questions.

How to use this book

The practice papers can be used in two main ways:

1. You can use the topic index on page 6 to find all the questions in the book that deal with a specific topic. This allows you to focus on areas you particularly wish to revise or, if you are only halfway through your course, it helps you to practice answering exam-style questions for the topics you have already studied.
2. If you have completed the course, you could try a complete practice paper in preparation for the final exam. You can do this in two ways:
   a) using your textbook and notes to help you.
   b) under exam style conditions, keeping a close eye on the clock. It's always useful to look at the total marks in the paper and the time available and make an estimate of how long a question should take. For example, you have $2\frac{1}{2}$ hours to complete the paper which has a total of 130 marks (20 marks in Section 1, 110 marks in section 2). This means that the multiple-choice questions in section 1 should take approximately 25 minutes to work through.

Revision advice

There's certainly plenty of revision advice available out there, so to keep things simple, we'd like to make just a few points:

1. There is good evidence to show that if you want to learn effectively and retain material you should **spread out your revision** and **revise on a regular basis**. This avoids having to re-learn everything you've previously studied. Last-minute cramming is a seriously risky business.

   Try to **review work** done soon after the lesson and then go back to the same material a few more times in the following weeks and months. It's okay to leave bigger time gaps between your reviews as time goes on.

2. When you review work done in class, staring at your textbook or your notes for long periods is largely a waste of time and highlighting text with lots of pretty colours may not make much difference either. Even making neater copies of your existing notes may not be worth it. Try making 'flash cards' with relevant facts/worked examples and use them whenever you have a spare moment.

   What you should do after reading a book or a section of your notes is put them aside and **test yourself**.

   Try some relevant **problems**.

   Or imagine that you have to **teach a class** the material you have just read. That should really make you think about what you have just been studying.

   Write down a few important points for your "lesson" (equations, main ideas) on a scrap of paper then teach them to your imaginary class. **Get active!**

3. Sticking to a weekly **timetable** for homework and revision is a pretty difficult job but organising your time can help you to get the work done more effectively and allow you to enjoy your free time without feeling guilty.

   In the beginning, sitting down to study is always a daunting task; however, provided you persevere, it does get easier and once you start getting answers to questions correct you will feel good about yourself and more confident in tackling new questions.

   Organisation is extremely important – avoid the temptation to put off to tomorrow what you can do today.

An example of a timetable appears below. As a Higher candidate, you can probably work for about half an hour at a time but if you're 'on a roll' it's perfectly reasonable to keep going for an hour. Make sure you take 10 to 15 minute breaks for a drink or to stretch your legs or listen to some music. (The jury is out on whether music while you're working helps you to study; some academics reckon that orchestral music is okay but vocal music is a distraction. You decide!)

Day	*6 pm–6·45 pm*	*7 pm–8 pm*	*8·15 pm–9 pm*	*9·15 pm–10 pm*
Monday	Homework	Homework	English revision	Maths revision
Tuesday	Maths revision	Physics revision	Homework	Free
Wednesday	Homework	Homework	English revision	French revision
Thursday	Homework	Maths Revision	Maths revision	Free
Friday	Geography revision	Physics Revision	Free	Free
Saturday	Free	Free	Free	Free
Sunday	Homework	Maths revision	French revision	Homework

In the Exam

Read the questions thoroughly before you attempt an answer. Make sure you understand what is being asked.

Questions often involve the use of a formula. Remember that **formulae** are provided in the SQA booklet. By the time of the final exam these formulae should also be planted in your memory!

You may need to use **data**, e.g. speed of light, $c = 3 \times 10^8 \text{ ms}^{-1}$. This may not appear in the question but can be found at the front of the question paper.

We wish you every success with Higher Physics.

Topic Index

	Practice Exam A		Practice Exam B		Practice Exam C	
Topics	***Section 1***	***Section 2***	***Section 1***	***Section 2***	***Section 1***	***Section 2***
Motion & graphs of motion	1, 3	2	1, 2	1	1, 2	1
Force, energy & power	5	1, 2	4	2, 3, 10		2, 12
Momentum & impulse	4	3, 4	3		3	3
Projectile motion	2					1
Gravitation	6	8	7		4	3
Special relativity	7	5	5	8	5	4
The Doppler effect	8		6		6	8
Hubble's Law			8	4		
The Big Bang				3		
The standard model	9			8	11	5
Forces on charged particles	11	6	9		13, 15	6
Nuclear reactions			10			7
The photoelectric effect	10, 12	7	19		7, 12	
Interference & diffraction	13, 15	9	12	7		9
Refraction of light	14		13	6	8, 9	10
Spectra & model of the atom	16		11, 14	9	10	
Measuring and monitoring alternating current	17		18		18	

	Practice Exam A		**Practice Exam B**		**Practice Exam C**	
Topics	***Section 1***	***Section 2***	***Section 1***	***Section 2***	***Section 1***	***Section 2***
Circuits: current, voltage, power & resistance			16		14, 19	12
Electrical sources & internal resistance	18		17		16, 17	
Capacitance		11		5		
Conductors, insulators & semiconductors	19	12	15			12
PN-junctions	20	12			20	
Uncertainties		10	20	10	3	

Practice Exam A

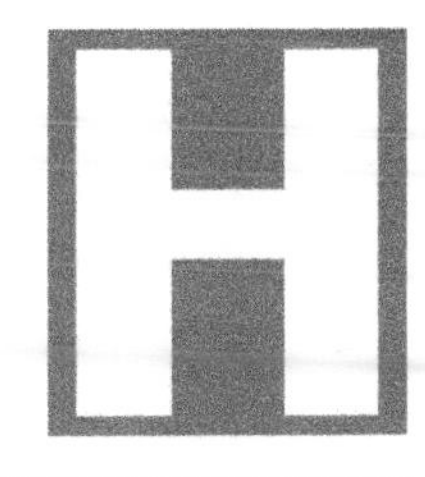

Higher Physics

Practice Papers for SQA Exams

Physics Section 1

Fill in these boxes:

Name of centre

Town

Forename(s)

Surname

Try to answer all of the questions in the time allowed.

Total marks – 130

Section 1 – 20 marks

Section 2 – 110 marks

Read all questions carefully before attempting.

You have $2\frac{1}{2}$ hours to complete this paper.

Write your answers in the spaces provided, including all of your working.

SECTION 1 ANSWER GRID

Mark the correct answer as shown ✓

	A	B	C	D	E
1	●	○	○	○	○
2	○	○	○	○	○
3	●	○	○	○	○
4	○	○	○	○	○
5	○	○	○	○	○
6	○	○	○	○	○
7	○	○	○	○	○
8	○	○	○	○	○
9	○	○	○	○	○
10	○	○	○	○	○
11	○	○	○	○	○
12	○	○	○	○	○
13	○	○	○	○	○
14	○	○	○	○	○
15	○	○	○	○	○
16	○	○	○	○	○
17	○	○	○	○	○
18	○	○	○	○	○
19	○	○	○	○	○
20	○	○	○	○	○

SECTION 1

1. A student records the velocity of a motorised cart along a level floor over a 10 second interval.

Time / s	0·0	2·0	4·0	6·0	8·0	10·0
Velocity / ms⁻¹	0·0	1·0	2·4	2·0	1·2	0·4

Which graph shows how the acceleration, *'a'* of the cart varies with time, *'t'* as it moves along the floor?

A

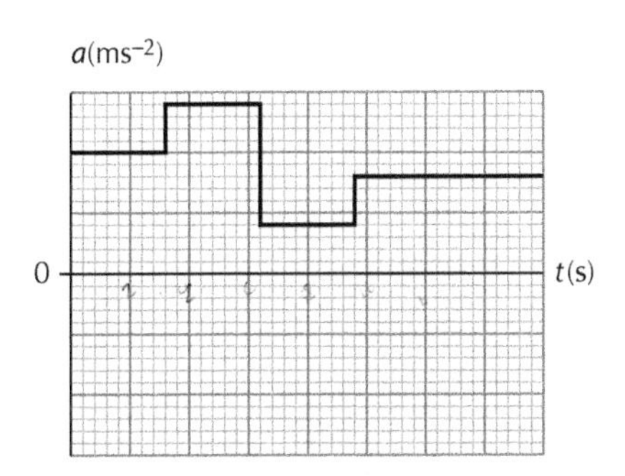

D

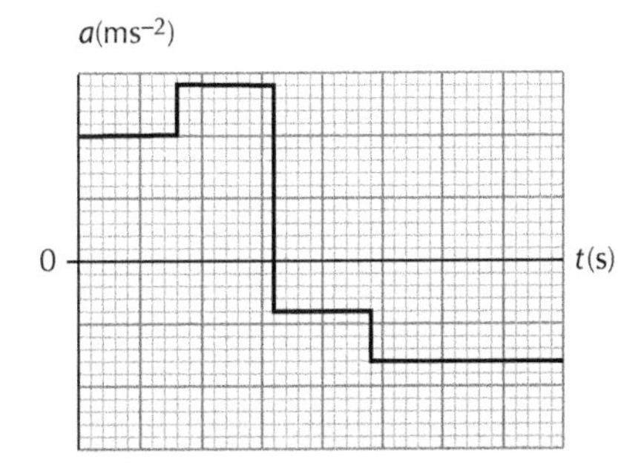

B

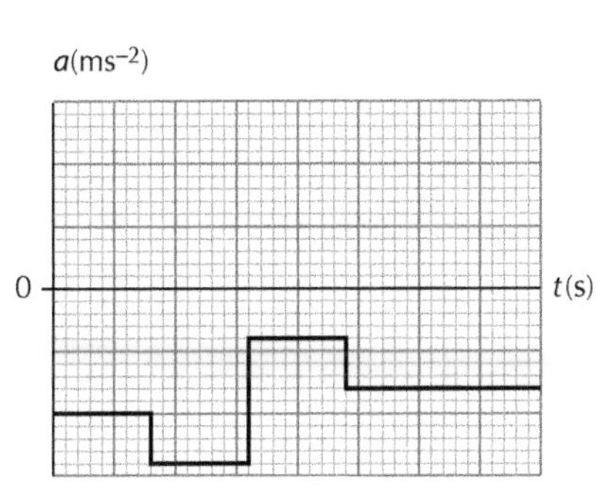

E

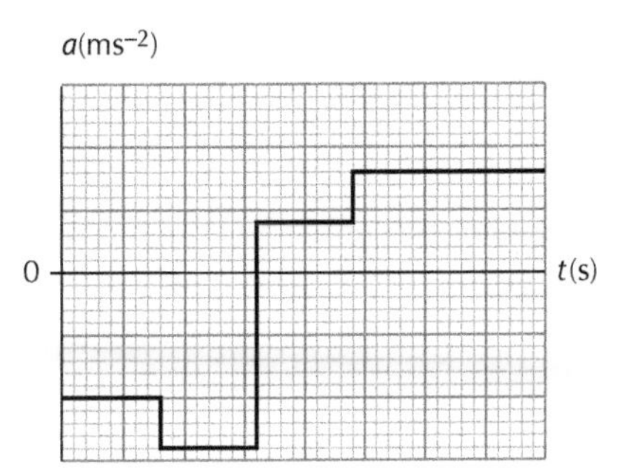

C

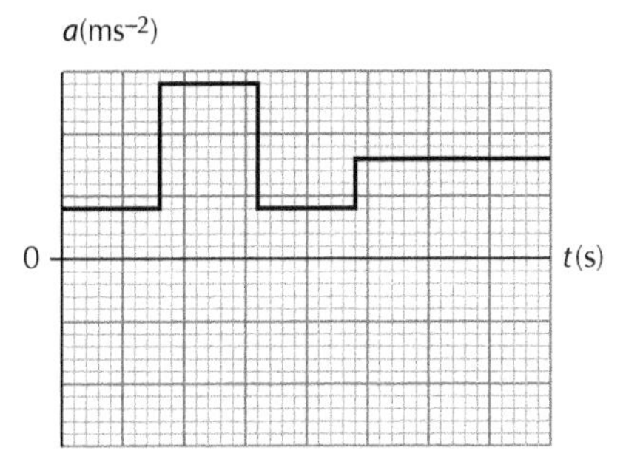

2. A ball is fired vertically into the air from a projectile launcher moving horizontally at constant speed.

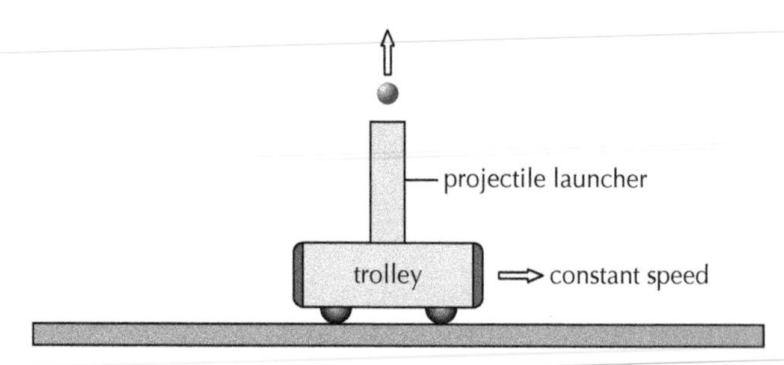

A special camera, running alongside the projectile launcher at the same constant speed as the trolley, takes strobe photographs at equal time intervals after the ball is fired until it reaches its highest position. Which image, produced by the camera, is shown?

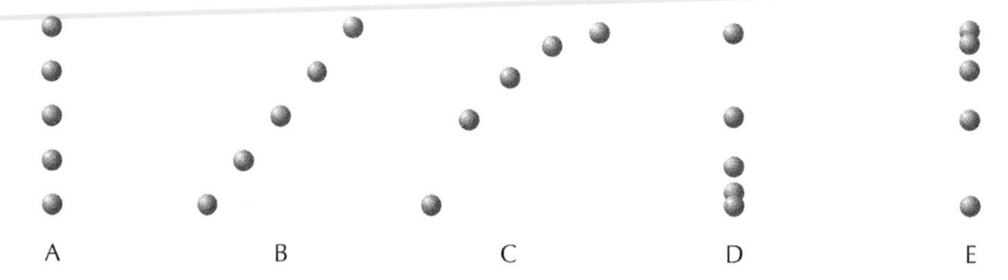

3. A ball bearing is projected along a bench and then onto a ramp. It rolls up and over the top of the ramp. The ramp has a rough surface on the way up and a smoother surface on the way down.

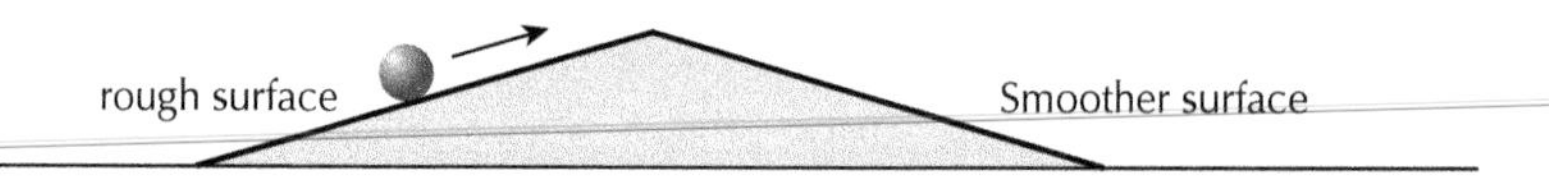

Which of the following velocity-time graphs represents the motion of the ball bearing on the two slopes?

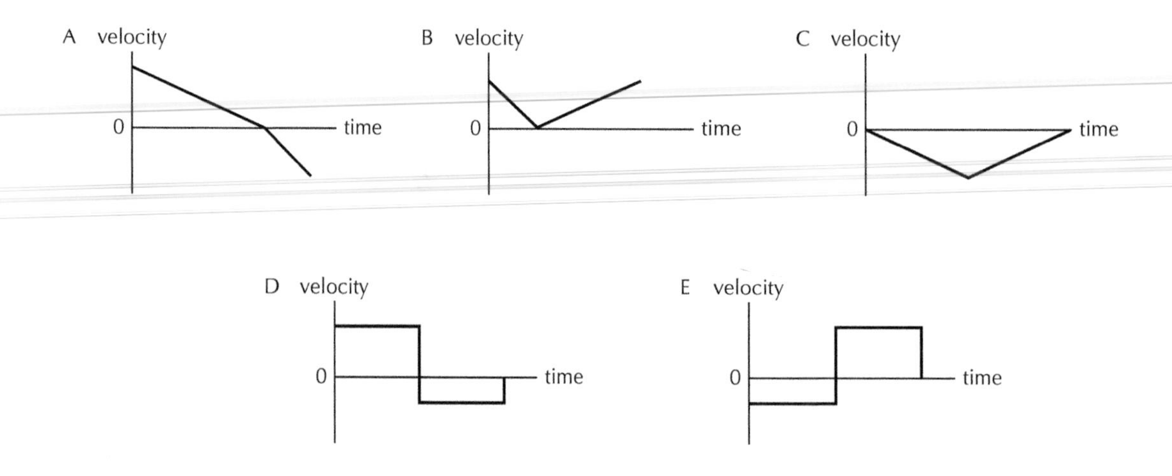

4. Baking soda crystals are used to sandblast a wall in order to remove paint.

A mass of 1000 kg of crystals strikes the wall horizontally each minute with a speed of 30 ms^{-1}.

If the crystals fall vertically after impact, the average force exerted upon the wall by the crystals is

A 500 N

B 1,800 N

C 2,000 N

D 9,800 N

E 30,000 N.

5. A train engine is pulling three carriages along a straight level track. The mass of the train engine is 50,000 kg and the mass of each carriage is 10,000 kg.

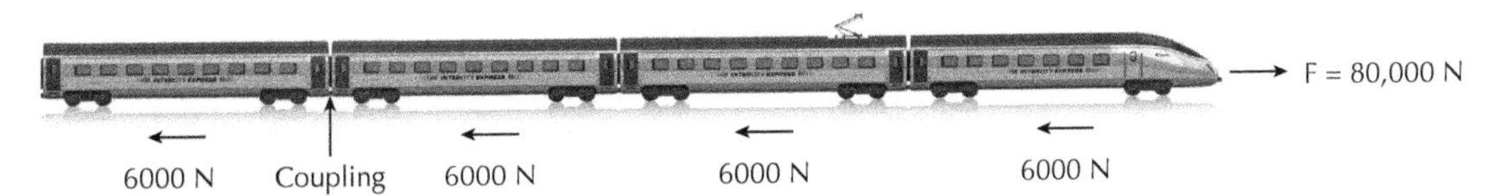

A frictional force of 6,000 N acts on the train engine and each carriage. The tension in the coupling between the second and third carriage is

A 7,000 N

B 10,000 N

C 13,000 N

D 16,000 N

E 26,000 N.

6. The mean distance of the planet Jupiter from the Sun is $7{\cdot}7 \times 10^8$ km. Data relating to the Sun and Jupiter is shown in the following table:

	Mass/kg	***Diameter/km***
Sun	$1{\cdot}99 \times 10^{30}$	$1{\cdot}39 \times 10^6$
Jupiter	$1{\cdot}90 \times 10^{27}$	$1{\cdot}43 \times 10^5$

The gravitational force acting on the planet Jupiter due to the Sun is

A $4{\cdot}25 \times 10^{23}$ N

B $4{\cdot}25 \times 10^{29}$ N

C $3{\cdot}24 \times 10^{35}$ N

D $3{\cdot}24 \times 10^{36}$ N

E $3{\cdot}24 \times 10^{38}$ N.

7. A spaceship is travelling with a constant speed of 0·7c past the Earth. The spaceship carries an identical clock to one at the launch site on Earth.

According to a stationary observer on Earth, which row in the table shows how the length of time for the journey shown on the spaceship's clock compares with the time on Earth and also how the length of the spaceship compares to its length at rest on Earth.

	Length of time	***Length of spaceship***
A	greater	longer
B	greater	shorter
C	smaller	longer
D	smaller	shorter
E	same	same

8. A truck is travelling at a constant speed along a straight stretch of motorway.

The driver of the truck sounds the truck horn as it approaches the bridge. The frequency of the sound emitted by the horn is 248 Hz.

An observer monitoring traffic noise from the bridge records a horn frequency of 268 Hz.

The speed of sound in air is 340 ms^{-1}.

The speed of the truck as it sounds its horn is approximately

A 16 ms^{-1}

B 20 ms^{-1}

C 25 ms^{-1}

D 27 ms^{-1}

E 45 ms^{-1}.

9. The following is part of a passage about subatomic particles with some words missing.
___(i)___ are intermediate mass particles which are made up of a quark-antiquark pair. Three quark combinations are called ___(ii) ___. All baryons are ___(iii)___.

The missing words are:

	(i)	**(ii)**	**(iii)**
A	fermions	baryons	bosons
B	baryons	fermions	leptons
C	mesons	baryons	fermions
D	baryons	mesons	fermions
E	mesons	baryons	bosons

10. Electromagnetic radiation with a wavelength of 330 nm is incident on sodium metal which has a work function of $3{\cdot}648 \times 10^{-19}$ J causing it to release photoelectrons.

The maximum speed of the photoelectrons emitted is

A $5{\cdot}1 \times 10^{5}$ ms^{-1}

B $7{\cdot}2 \times 10^{5}$ ms^{-1}

C $8{\cdot}9 \times 10^{5}$ ms^{-1}

D $1{\cdot}5 \times 10^{6}$ ms^{-1}

E $5{\cdot}2 \times 10^{11}$ ms^{-1}.

11. A proton enters into a region of magnetic field pointing out of the page as shown:

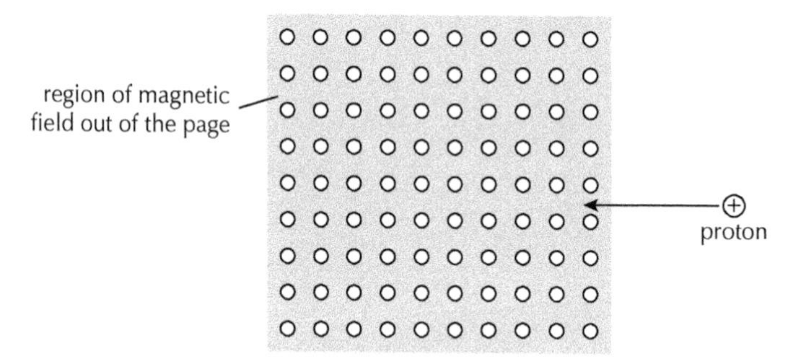

The magnetic field exerts a force upon the proton making it move

A to the right

B out of the page

C into the page

D towards the bottom of the page

E towards the top of the page.

12. A student makes three statements about wave particle duality:

I Line spectra provide evidence for the particulate nature of light.

II The threshold wavelength is the maximum wavelength of a photon required for photoemission.

III The work function of a metal is the minimum energy required for photoemission.

Which of these statements is/are true?

A I only

B II only

C III only

D I and II only

E II and III only

13. A 900 lines per mm diffraction grating is used with a 500 nm monochromatic light source to observe constructive interference fringes on a screen.

The maximum number of possible fringes is

A 2

B 4

C 5

D 8

E 9.

14. Which row in the following table describes what happens to the wavespeed, wavelength and amplitude of a ray of monochromatic light as it travels from water into air?

	wavespeed	***wavelength***	***amplitude***
A	increases	decreases	unchanged
B	decreases	decreases	increases
C	unchanged	increases	decreases
D	decreases	unchanged	increases
E	increases	increases	unchanged

15. Two identical loudspeakers are used to demonstrate sound interference. They are connected to the same signal generator output terminals as shown.

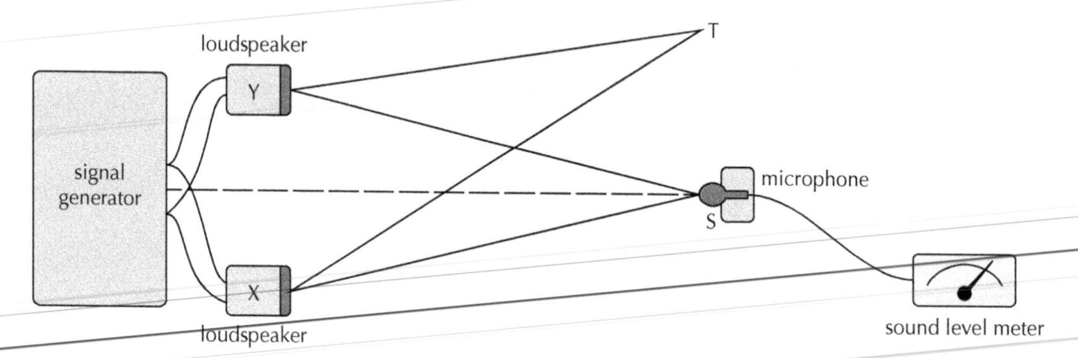

Position S is the same distance from both loudspeakers and the sound level meter indicates a maximum reading.

As the microphone moves towards position T, the sound level meter indicates alternate minimum and maximum readings.

T corresponds to the third minimum reading. The distance from X to T is 4·31 m and the distance from Y to T is 4·01 m.

The speed of sound is 340 ms^{-1}.

The frequency of the sound produced by the speakers is

A 1130 Hz

B 2270 Hz

C 2830 Hz

D 3400 Hz

E 3970 Hz.

16. A textbook shows transitions between energy levels in an atom. Which energy level diagram shows the transition made by an electron producing a photon with the longest wavelength?

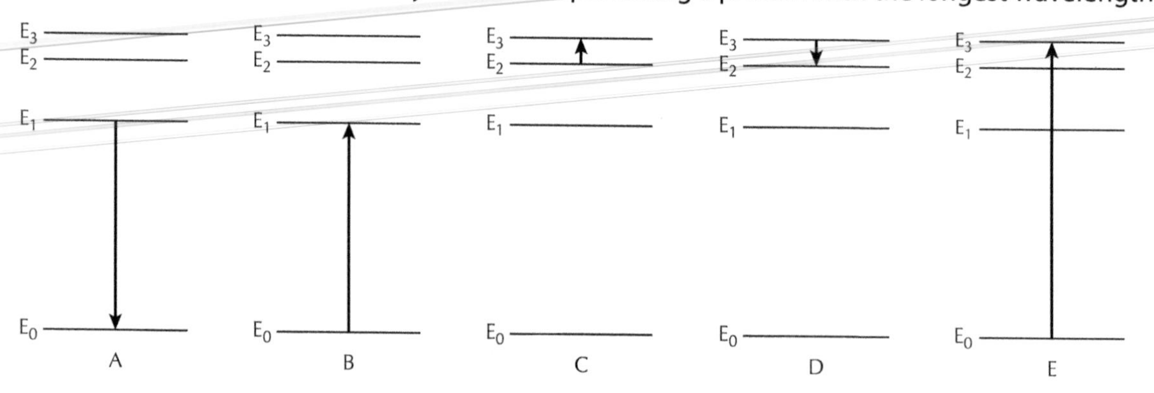

17. An oscilloscope is used to display an alternating voltage. The Y-gain and timebase settings are shown.

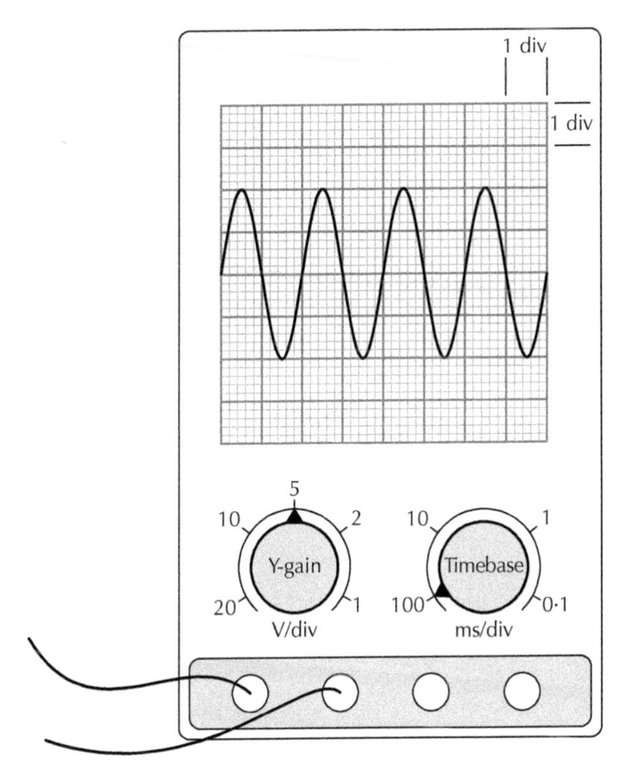

Which row in the table shows the values of the peak voltage and frequency of the a.c. signal?

	Peak Voltage/V	***Frequency/Hz***
A	10	4
B	10	5
C	10	200
D	20	5
E	20	200

18. Four identical cells each with an e.m.f. of 1·5 V and internal resistance 0·20 Ω are connected in parallel across a high resistance voltmeter as shown.

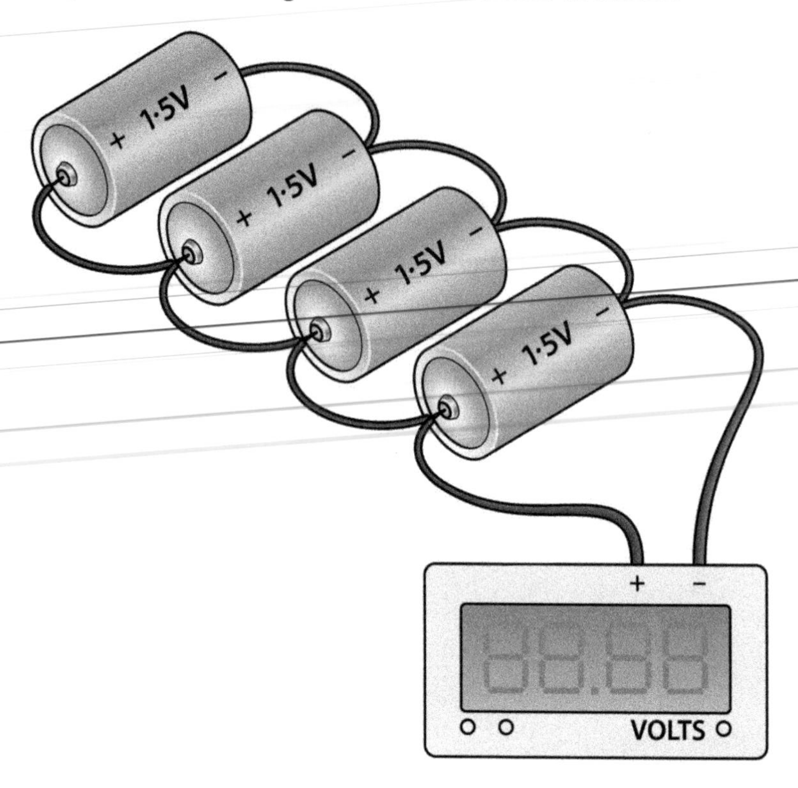

Which row in the table shows the overall e.m.f and internal resistance of this combination of cells?

	e.m.f / V	*internal resistance / Ω*
A	1·5	0·05
B	1·5	0·20
C	1·5	0·80
D	6·0	0·05
E	6·0	0·80

19. A crystal of silicon has some of its atoms replaced by gallium atoms in a process called "doping".

Gallium atoms have only three outer electrons whereas silicon atoms have four outer electrons.

This doping procedure will

A create an n-type semiconductor

B decrease the resistance of the crystal due to the creation of holes

C increase the resistance of the crystal due to the creation of holes

D decrease the resistance of the crystal due to the creation of electrons

E increase the resistance of the crystal due to the creation of electrons.

20. A student writes the following three statements about band theory in a notebook:

I In a conductor there is no band gap between the valence band and the conduction band.

II In a semiconductor the band gap between the valence band and the conduction band is larger than the band gap in an insulator.

III In an insulator the conduction band is full.

Which of these statements is/are true?

A I only

B II only

C III only

D I and II only

E II and III only

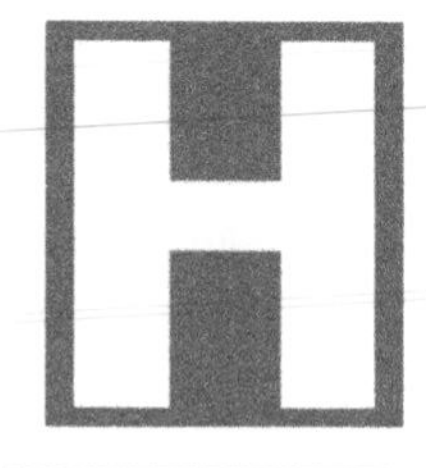

Higher Physics

Practice Papers for SQA Exams

Physics Section 2

Fill in these boxes:

Name of centre

Town

Forename(s)

Surname

Try to answer all of the questions in the time allowed.

Total marks – 130

Section 1 – 20 marks

Section 2 – 110 marks

Read all questions carefully before attempting.

You have $2\frac{1}{2}$ hours to complete this paper.

Write your answers in the spaces provided, including all of your working.

SECTION 2

1. A cart with a mass of 0·40 kg accelerates up a slope at 0·80 ms^{-2}.

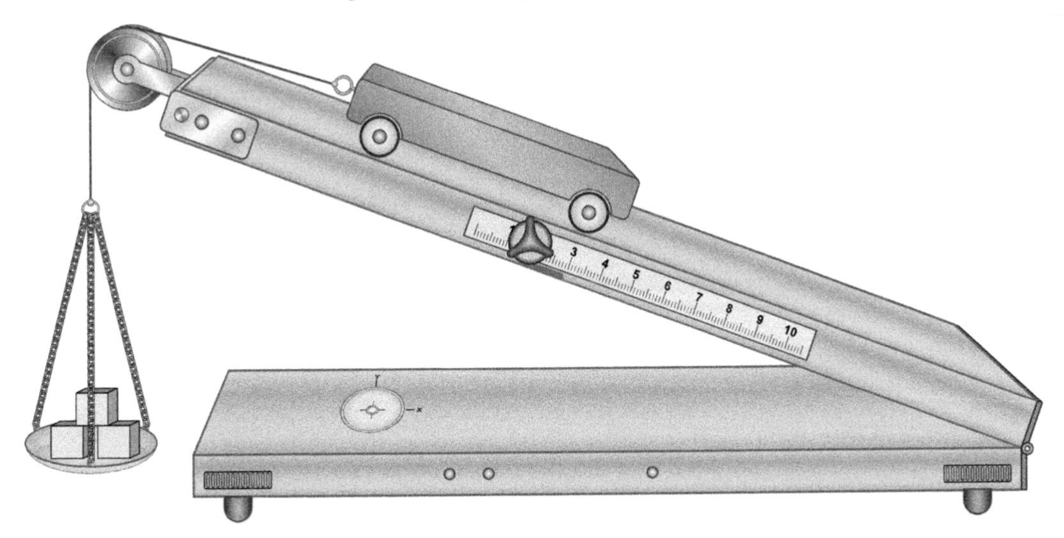

A forward force of 1·98 N is provided by a falling mass carrier and masses attached to the cart via a frictionless pulley and a lightweight cord. A frictional force of 0·20 N acts down the incline.

(a) Calculate the magnitude of the unbalanced force acting on the cart. **3**

Space for working and answer

(b) Calculate the angle of inclination that the slope makes with the horizontal bench on which it sits. **5**

Space for working and answer

MARKS

DO NOT WRITE IN THIS MARGIN

(continued)

(c) Assuming that the frictional force remains the same at any angle of inclination, calculate the angle required for the cart to travel up the slope at constant speed. You may assume the forward force is still 1·98 N. **4**

Space for working and answer

Total marks 12

2. A helicopter is transporting supplies to a remote area. The helicopter has a mass of 3850 kg and the supplies have a mass of 650 kg.

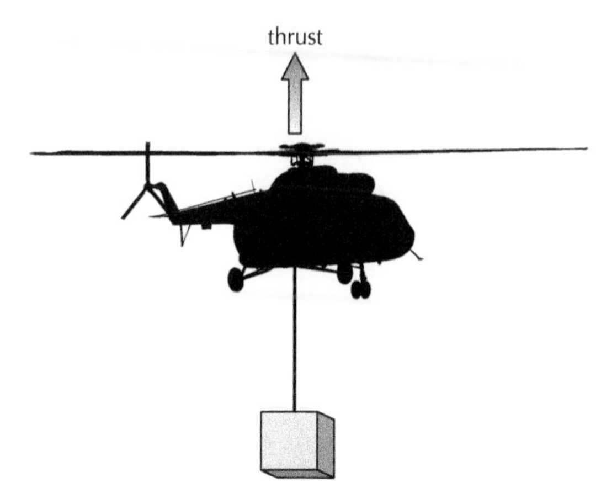

(a) Calculate the upward thrust from the helicopter blades required to accelerate the helicopter and the supplies upwards at 2 ms^{-2}. **3**

Space for working and answer

(continued)

(b) Once it has achieved a safe height above the surrounding trees the helicopter moves off horizontally at a steady speed exerting a horizontal force of 3678 N on the supplies.

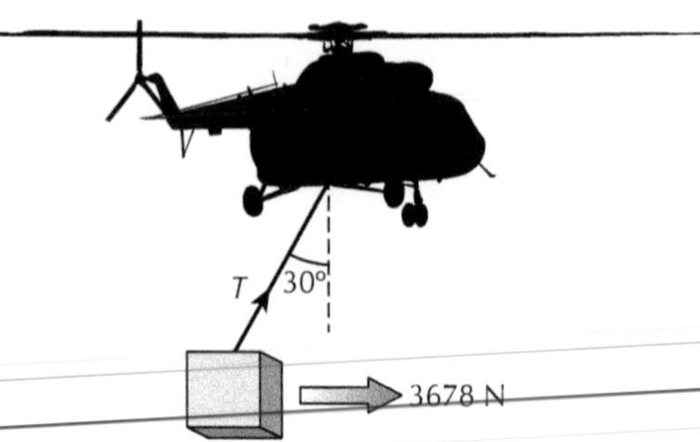

Calculate the tension, T, in the cable attaching the supplies to the helicopter if it makes an angle of 30° to the vertical as shown. **4**

Space for working and answer

(c) After depositing the supplies the helicopter returns to its original location where it hovers for a few seconds at a height of 32 m.

(i) Calculate the upward thrust required from the helicopter blades whilst it hovers at this height. **3**

Space for working and answer

(continued)

(ii) The helicopter now descends from the height of 32 m with an initial speed of 2 ms^{-1}.
Show that the deceleration required to bring the helicopter to rest carefully on the ground is 0·06 ms^{-2}. **2**

Space for working and answer

(iii) Calculate the time taken for the helicopter to land. **3**

Space for working and answer

Total marks 15

3. At the Glasgow Commonwealth Games curlers used granite stones with a mass of 18·0 kg and 20·0 kg, which slid along an ice rink towards a target.

The aim of the game is to get the stone as close as possible to the centre of the target. Measurements made during one slide of an 18 kg stone allowed the frictional force between the stone and the ice to be recorded. A graph of the data is plotted as shown.

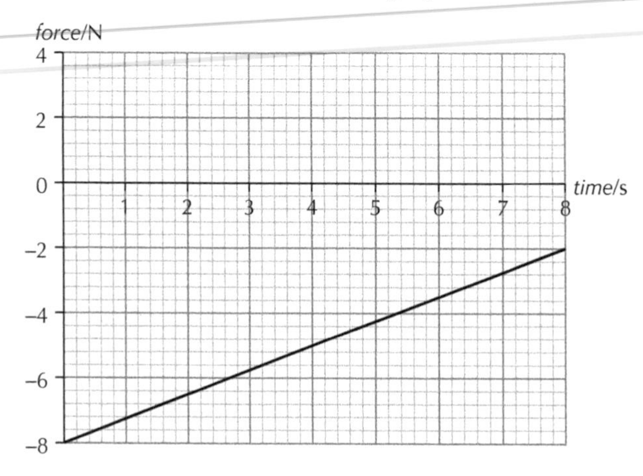

(a) (i) Calculate the change in momentum of the 18·0 kg stone over the first 4·0 seconds. **3**

Space for working and answer

(ii) Show that the change in velocity after 4·0 s is 1·44 ms^{-1}. **2**

Space for working and answer

(continued)

(b) Near the centre of the target the 18·0 kg stone encounters an opponent's stone which is stationary. The opponent's stone has a mass of 20·0 kg and when struck moves off in a straight line at 0·60 ms^{-1}.

(i) If the 18 kg stone continues in a straight line at 0·40 ms^{-1} after the collision, calculate its approach speed just before the collision. **3**

Space for working and answer

(ii) Is the collision elastic or inelastic?
You must justify your answer. **3**

Space for working and answer

MARKS | DO NOT WRITE IN THIS MARGIN

(continued)

(c) The collision lasted for 10 ms. Calculate the average force exerted upon the 20 kg stone during the collision. **3**

Space for working and answer

Total marks 14

4. In karate lessons students are often taught to terminate a strike by the fist a few centimetres inside the body of an opponent which is totally different from most other forms of combat which usually involves 'follow through' by the fist.

Use your knowledge of Physics to comment upon which technique will produce more damage to the opponent. **3**

Total marks 3

5. A binary star system observable from Earth consists of two stars orbiting around a common centre of mass. The two stars have the same period and are always opposite each other as shown.

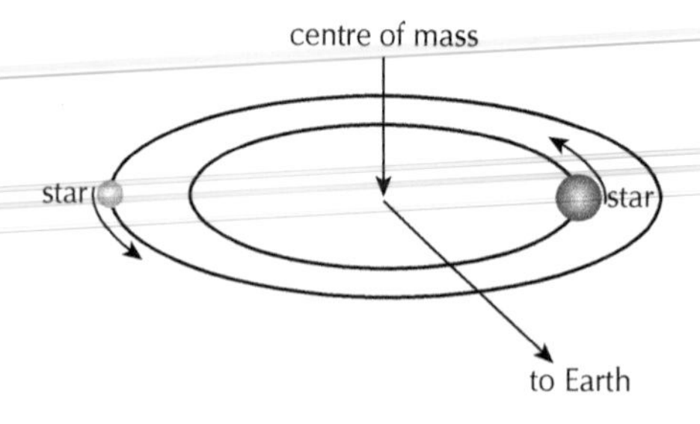

The binary system and the Earth lie in the same plane. The wavelength of a particular emission line in the smaller star's spectrum varies from 655.96 nm to 665.60 nm.

MARKS | DO NOT WRITE IN THIS MARGIN

(continued)

(a) Explain why the line observed varies in wavelength. **2**

(b) (i) Calculate the maximum recessional velocity of the smaller star relative to the Earth. **4**

Space for working and answer

(b) (ii) Another line in the spectrum emitted by the smaller star has a wavelength of 434.05 nm.

What is the maximum wavelength of this line when viewed from the Earth? **3**

Space for working and answer

Total marks 9

6. A cyclotron is a particle accelerator used to accelerate charged particles used in high-energy collisions.

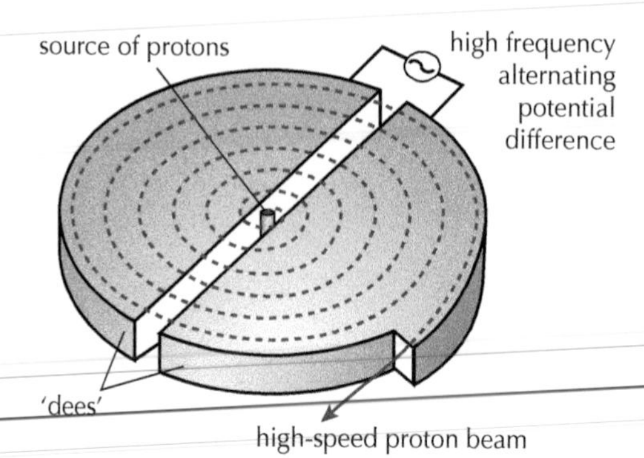

The acceleration is provided by an alternating potential difference across the gap between the 'dees' and the spiral path is controlled by a constant magnetic field acting perpendicularly to the path taken by the particles.

(a) Explain why the potential difference across the gap has to be an *alternating potential difference.* **3**

(b) The alternating potential difference used in one particular cyclotron is 28 kV. A proton starts at rest near the centre of the cyclotron and accelerates across the gap when the electric field is applied.

Calculate the increase in speed as the proton crosses the gap. **3**

Space for working and answer

(continued)

(c) The time required for a proton to move in a semi-circular path around one of the 'dees' is given by:

$$t = \frac{\pi m}{qB}$$

where m = the mass of the proton;
q = the charge of the proton, and
B = the magnetic field strength.

Use this equation to explain why the proton travels in an increasing spiral path. (You may assume that the maximum speed of the protons is not greater than 10% of the speed of light.) **3**

(d) Suggest 2 reasons why the cyclotron does not use electrons for high-energy collisions with the nucleus of an atom within a metal target. **2**

Total marks 11

7. Calcium metal has a work function of $4{\cdot}64 \times 10^{-19}$ J. A strip of calcium is illuminated with an ultraviolet light source of wavelength 340 nm.

(a) (i) Show that the threshold frequency for the calcium metal used is 7×10^{14} Hz. **2**

Space for working and answer

(continued)

(ii) Calculate the maximum velocity of the photoelectrons released. **4**

Space for working and answer

(b) The irradiance of the ultraviolet light source used is now reduced. Describe and explain the impact this has on (i) the number of photoelectrons and (ii) the maximum velocity of the photoelectrons liberated. **3**

Total marks 9

8. The Sun and the Earth both exert a force of attraction on the Moon. A student suggests that since the Moon is closer to the Earth then the Earth must exert a greater force on the Moon than the Sun which is much further away.

Use your knowledge of Physics to comment on this suggestion. **3**

Total marks 3

9. A single slit, a double slit and a monochromatic light source are used to investigate a similar interference experiment to that carried out by Thomas Young with white light in 1801.

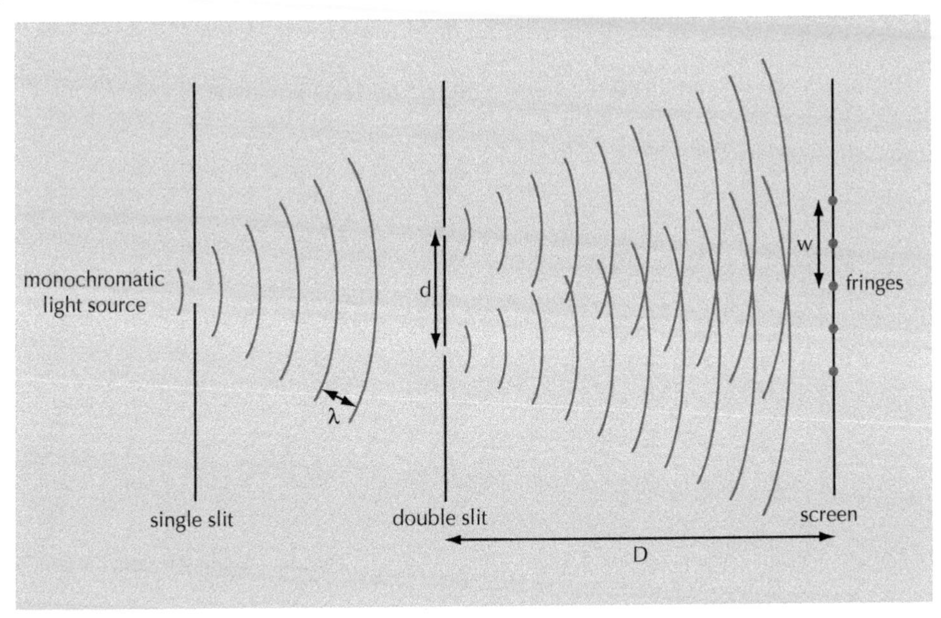

(a) Explain why a single slit is used in front of the double slit. **1**

The distance from the double slit to the screen (D) is measured 5 times using a hand-held ultrasonic device. The measurements obtained are:

(1·921 m, 1·927 m, 1·924 m, 1·922 m, 1·924 m)

(b) (i) Calculate the mean value of these readings together with the random uncertainty in this value. **2**

Space for working and answer

MARKS | DO NOT WRITE IN THIS MARGIN

(continued)

(ii) The expression

$$m\lambda D = dw$$

can be used to calculate the wavelength (λ) of the monochromatic light source.

If the distance (w) between the central fringe and the 2nd order maximum, (m) is $15{\cdot}50 \pm 0{\cdot}05$ mm and the distance between the slits (d) is $0{\cdot}126 \pm 0{\cdot}001$ mm, calculate the wavelength and absolute uncertainty of the monochromatic light source used.

Express your answer in the form (wavelength ± absolute uncertainty) in nm. **6**

Space for working and answer

(iii) What is the colour of this light source? **1**

(c) What would happen to the distance, w if the distance, D was increased? **1**

(d) The experiment is repeated with a white light source rather than a monochromatic light source.

Describe what would now be seen on the screen. **2**

Total marks 13

10. A student uses the circuit below together with a stopwatch to carry out an experiment to produce data which would allow a graph of the charge stored by a capacitor against the p.d. across the capacitor to be plotted.

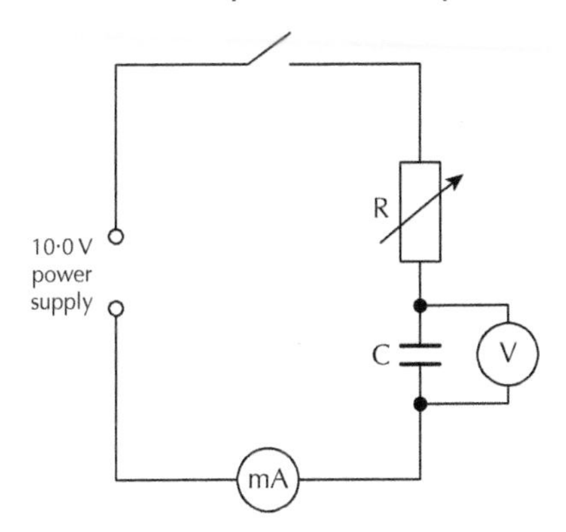

The data obtained from the experiment is shown in the following table:

Time / s	5·0	10·0	15·0	20·0	25·0
P.d. / V	2·0	4·0	6·0	8·0	10·0
Current / mA	4·0	4·2	3·9	4·0	4·0

(a) Copy the table and add a row to show the charge stored by the capacitor for each applied p.d. **2**

(b) Describe how the experiment is carried out to obtain the charge stored by the capacitor for each p.d. across the capacitor **3**

(c) Using the graph paper provided at the back of this book plot a graph of *charge* (mC) against *p.d.* (V). **2**

MARKS | DO NOT WRITE IN THIS MARGIN

(continued)

(d) Determine how much energy is stored by the capacitor when the p.d. across it is 10·0 V? **3**

Space for working and answer

(e) Show that when the voltage across the capacitor is 5·4 V, the energy stored by the capacitor is 146 mJ. **4**

Space for working and answer

Total marks 14

11. A company manufactures silicon wafers which are used in the production of semiconductors used in the telecommunication industry.

The silicon wafers are doped with small amounts of arsenic in order to produce the semiconductors.

(a) (i) What are the majority charge carriers in this type of semiconductor? You must explain your answer. **2**

(ii) What charge does the doped silicon have? You must explain your answer. **2**

(continued)

(b) The semiconductor formed is used to produce a p-n junction diode. Explain using the terms, *valence band* and *conduction band* how this diode conducts when placed in forward bias mode. **3**

Total marks 7

[END OF QUESTION PAPER]

Practice Exam B

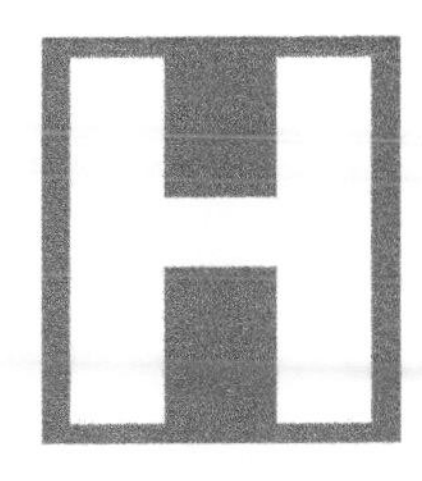

Higher Physics

Practice Papers for SQA Exams

Physics Section 1

Fill in these boxes:

Name of centre

Town

Forename(s)

Surname

Try to answer all of the questions in the time allowed.

Total marks – 130

Section 1 – 20 marks

Section 2 – 110 marks

Read all questions carefully before attempting.

You have $2\frac{1}{2}$ hours to complete this paper.

Write your answers in the spaces provided, including all of your working.

SECTION 1 ANSWER GRID

Mark the correct answer as shown ✓

	A	B	C	D	E
1	○	○	○	○	○
2	○	○	○	○	○
3	○	○	○	○	○
4	○	○	○	○	○
5	○	○	○	○	○
6	○	○	○	○	○
7	○	○	○	○	○
8	○	○	○	○	○
9	○	○	○	○	○
10	○	○	○	○	○
11	○	○	○	○	○
12	○	○	○	○	○
13	○	○	○	○	○
14	○	○	○	○	○
15	○	○	○	○	○
16	○	○	○	○	○
17	○	○	○	○	○
18	○	○	○	○	○
19	○	○	○	○	○
20	○	○	○	○	○

SECTION 1

1. Forces are usually measured in newtons. An alternative unit of force is

A kgs^{-1}

B kgm^{-1}

C $kgms^{-1}$

D $kgms^{-2}$

E $kgm^{2}s^{-2}$

2. The displacement–time graph for a moving object is shown below.

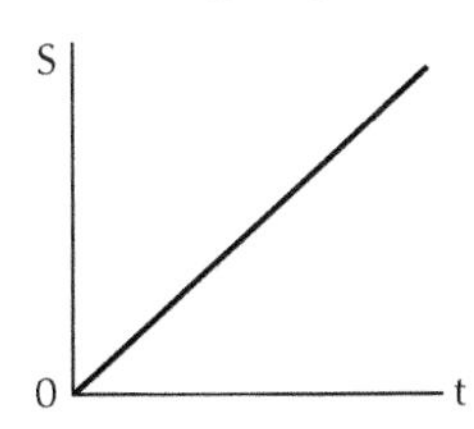

Which one of the following velocity–time graphs could represent the same motion?

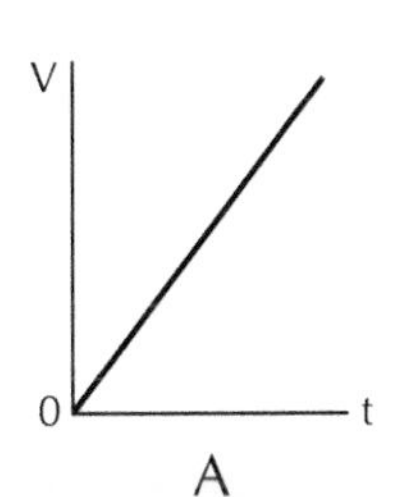

A

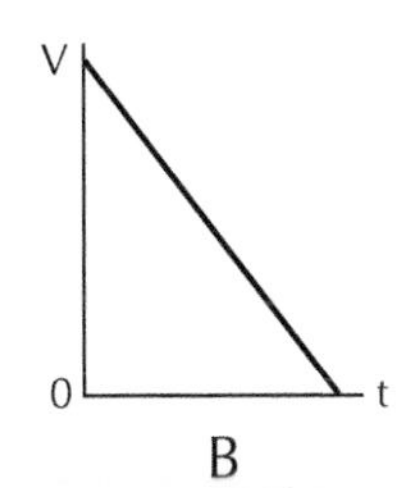

B

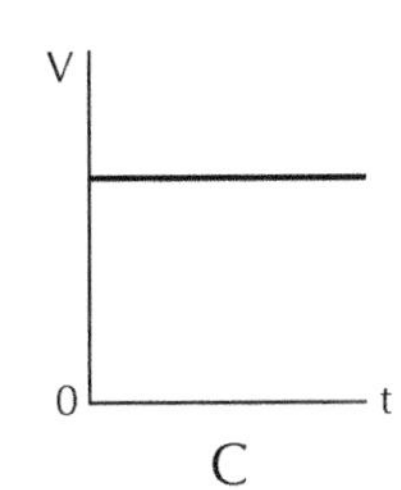

C

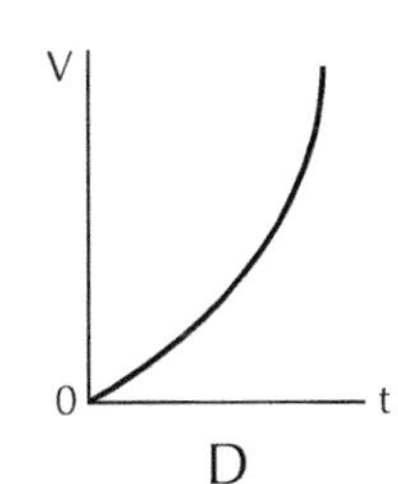

D

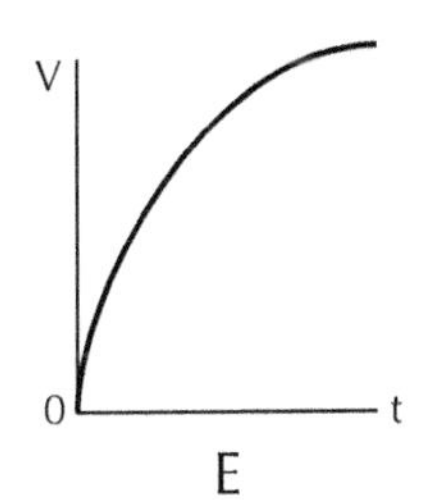

E

3. The graph shows how the force on a 2 kg object varies with time.

The impulse on the object is

A zero

B 20 Ns

C 40 Ns

D 60 Ns

E 120 Ns.

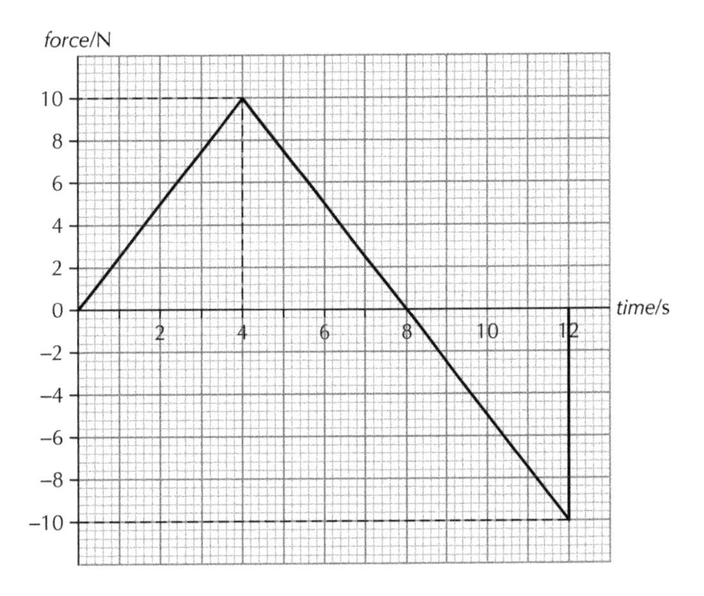

4. A 20 N force acts on the block of wood shown below.

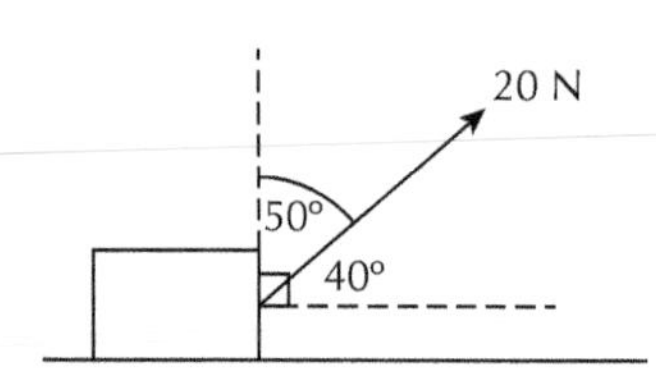

Which row in the table gives correct values for both the horizontal and vertical components of the force?

	Horizontal Component/N	***Vertical Component/N***
A	20 sin 50°	20 cos 40°
B	20 cos 40°	20 sin 40°
C	20 cos 40°	20 sin 50°
D	20 cos 50°	20 sin 50°
E	20 sin 40°	20 cos 40°

5. A 10 cm diameter meteorite travelling at a speed of 0·6 C passes a stationary observer. Relative to the observer, the diameter of the meteorite (in its direction of travel) will be

A 6·4 cm

B 8·0 cm

C 10 cm

D 12·5 cm

E 1·0 cm.

6. A police car siren has a frequency of 960 Hz. The car approaches a pedestrian at a speed of 40 ms^{-1}. If the speed of sound in the air is 340 ms^{-1}, the frequency of the sound heard by the pedestrian will be

A 1088 Hz

B 1073 Hz

C 960 Hz

D 859 Hz

E 847 Hz.

7. It is estimated that the comet Hale–Bopp will return to our solar system around the year 4385. The comet which has a mass of $1{\cdot}3 \times 10^{16}$ kg will pass closest to Earth which has a mass of $6{\cdot}0 \times 10^{24}$ kg, at a distance of 150 million km. At that point the gravitational force of the comet on our planet is

A 0 N

B 9·8 N

C $2{\cdot}3 \times 10^{8}$ N

D $3{\cdot}5 \times 10^{19}$ N

E $1{\cdot}2 \times 10^{39}$ N.

8. The wavelength of a line in the calcium emission spectrum is 393·3 nm when measured in a laboratory. The wavelength of the same line emitted from a distant galaxy is 396·2 nm. This means that the galaxy is

A approaching Earth at 3×10^{8} ms^{-1}

B approaching Earth at $2{\cdot}2 \times 10^{6}$ ms^{-1}

C moving away from Earth at 3×10^{8} ms^{-1}

D moving away from Earth at $2{\cdot}2 \times 10^{6}$ ms^{-1}

E moving away from Earth at $2{\cdot}3 \times 10^{-18}$ ms^{-1}.

9.

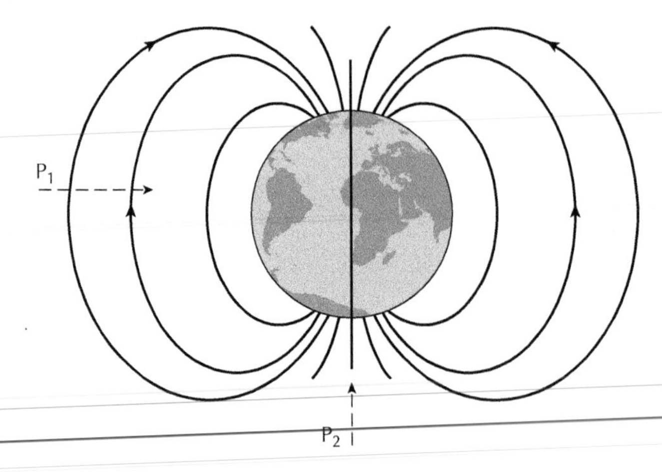

The diagram shows the paths of two cosmic ray protons P_1 and P_2 entering the Earth's magnetic field at the same speed. Which statement correctly describes the effect of the Earth's field on the two protons?

A The force on both P_1 and P_2 is zero.

B The forces on P_1 and P_2 are the same in both size and direction.

C The forces on P_1 and P_2 are the same size but in opposite directions.

D The force on P_1 is very much larger than the force on P_2.

E The force on P_2 is very much larger that the force on P_1.

10. The radioactive nuclide Thorium 232 decays into one of its isotopes Thorium 228. The particle(s) emitted during this process is/are

A one α particle

B one β particle

C two α particles and one β particle

D two α particles and two β particles

E one α particle and two β particles.

11. The diagram shows three energy levels for electrons in an atom. E_0 is the lowest level.

If transition E_0 to E_1 represents absorption of green light, the transition E_2 to E_0 could represent

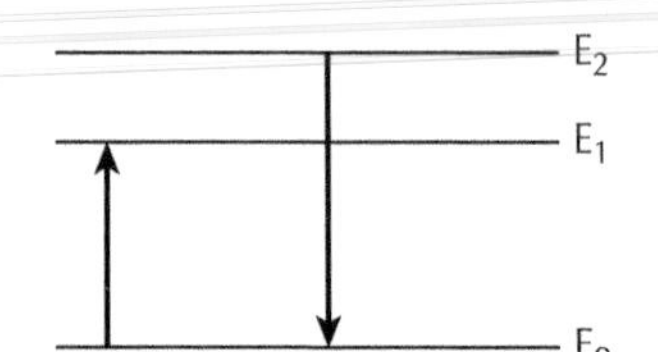

A absorption of blue light

B absorption of red light

C emission of blue light

D emission of red light

E emission of infra red.

12. A diffraction pattern is produced when monochromatic light passes through a grating. The first order maximum is produced at angle θ.

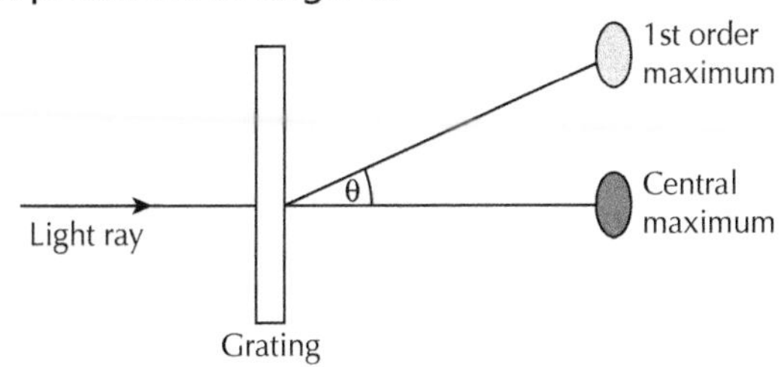

The wavelength of the light and the number of lines per mm required to produce the smallest angle is:

	Wavelength/nm	***Grating lines per mm***
A	400	300
B	400	600
C	500	300
D	600	600
E	700	600

13. The diagram shows a ray of monochromatic light passing through a semicircular glass block.

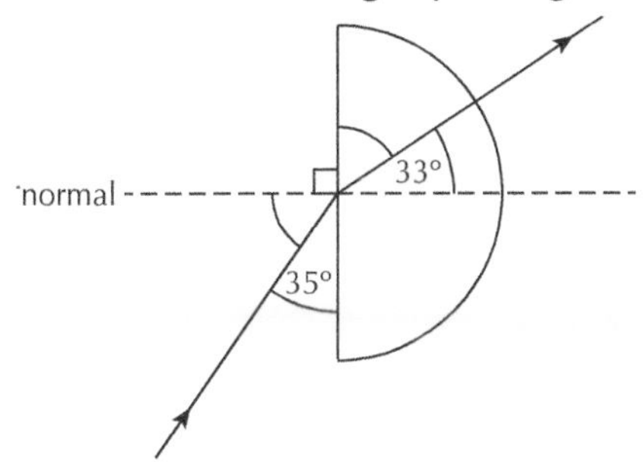

The refractive index of the glass is

A $\frac{\sin 35°}{\sin 33°}$

B $\frac{\sin 33°}{\sin 35°}$

C $\frac{\sin 33°}{\sin 55°}$

D $\frac{\sin 55°}{\sin 57°}$

E $\frac{\sin 55°}{\sin 33°}$

14. A very small lamp illuminates a picture on a wall 1 metre away with an irradiance I. The lamp is moved away to a distance of 2 metres. The new irradiance will be

A $\sqrt{2}I$

B $2I$

C $\frac{I}{\sqrt{2}}$

D $\frac{I}{2}$

E $\frac{I}{4}$

15. P type silicon is made when a few silicon atoms are replaced by boron atoms in a silicon crystal. The effect of this process on the crystal is to

A increase its resistance

B make it positively charged

C make it negatively charged

D reduce its resistance

E allow it to conduct only in one direction.

16. All the resistors shown below have the same resistance. Which combination has the biggest resistance between terminals X and Y?

A

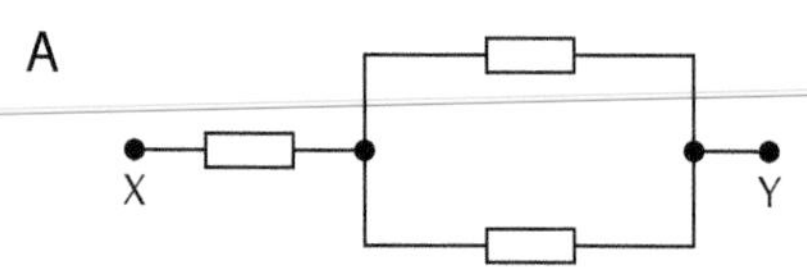

B

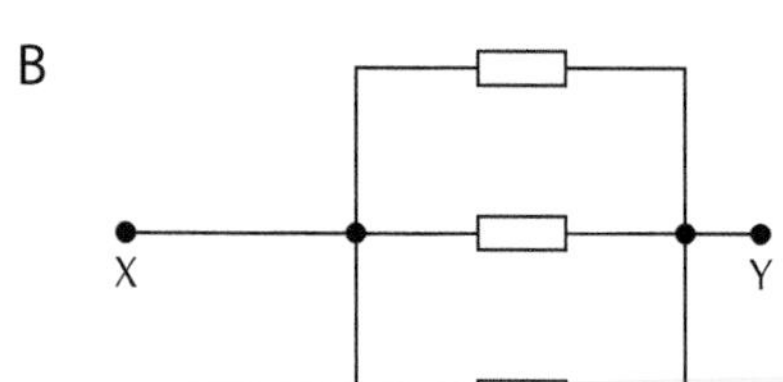

C

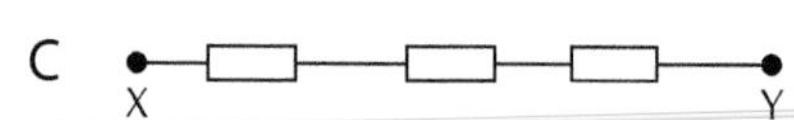

D

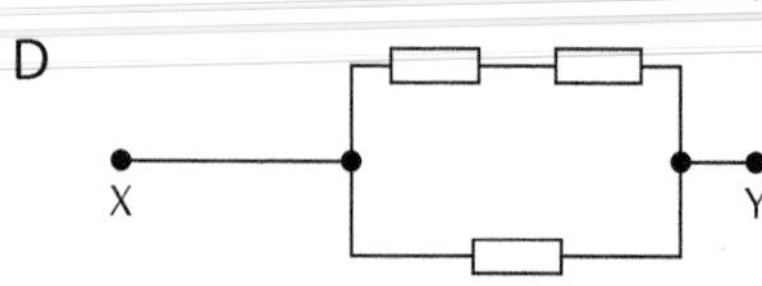

E

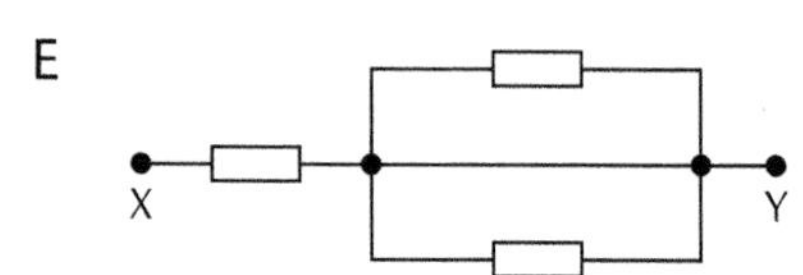

17. A cell with an emf of 1·6 V and internal resistance, r, is connected as shown below. When the switch is closed, the voltmeter reading is 1·2 V and the ammeter reading is

A 0·2 A

B 0·4 A

C 0·6 A

D 0·8 A

E 1·0 A.

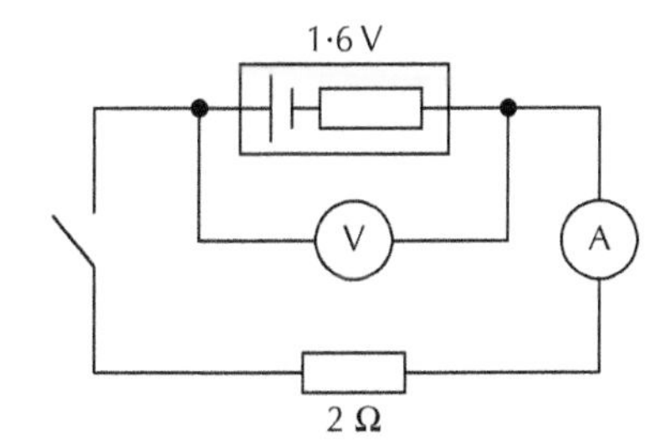

18. The waveform on this oscilloscope has a frequency of 50 Hz.

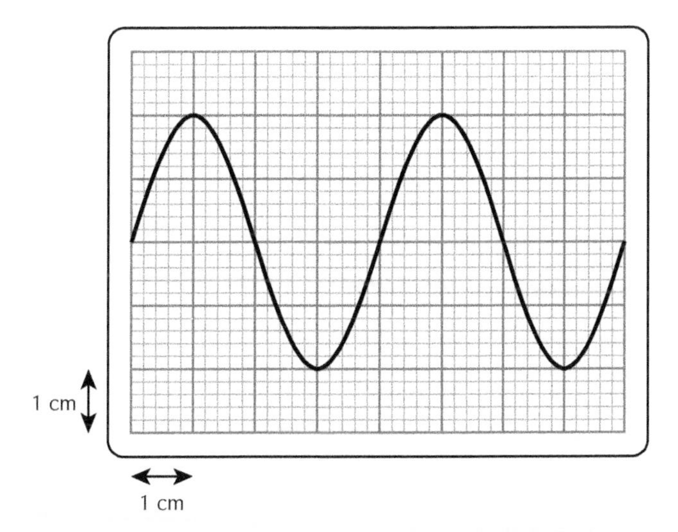

The timebase setting is

A 2·5 ms per cm

B 5·0 ms per cm

C 10 ms per cm

D 20 ms per cm

E 25 ms per cm.

19. A raybox with a violet filter is used to produce a photoelectric current in the circuit below.

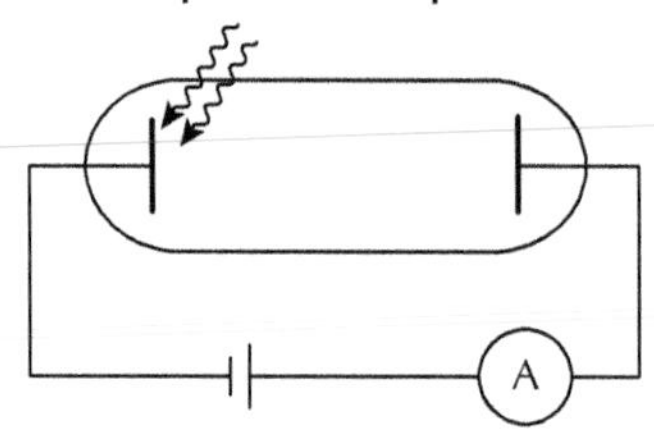

The irradiance of the raybox lamp is increased. Which row in the table correctly shows the effect of this increase on the ammeter reading and the kinetic energy of each photoelectron?

	Ammeter reading	*Kinetic energy of each photoelectron*
A	decreased	decreased
B	unchanged	unchanged
C	unchanged	decreased
D	increased	unchanged
E	increased	increased

20. A pupil uses the equipment shown below to make measurements of length, temperature and time.

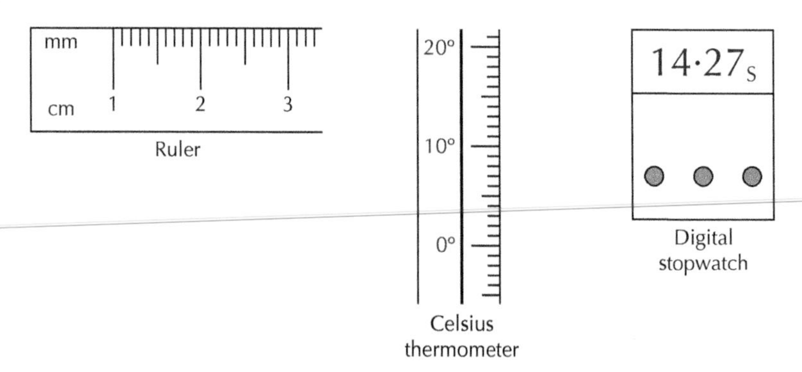

Which measurement would have the largest percentage scale reading uncertainty?

A A time of 1·00 s.

B A time of 10 s.

C A temperature of 5°C.

D A length of 3 cm.

E A length of 2 mm.

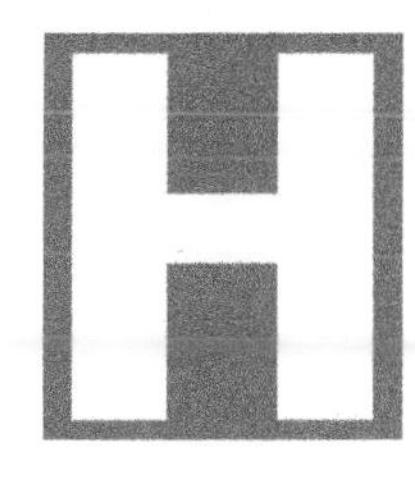

Higher Physics

Practice Papers for SQA Exams

Physics Section 2

Fill in these boxes:

Name of centre

Town

Forename(s)

Surname

Try to answer all of the questions in the time allowed.

Total marks – 130

Section 1 – 20 marks

Section 2 – 110 marks

Read all questions carefully before attempting.

You have $2\frac{1}{2}$ hours to complete this paper.

Write your answers in the spaces provided, including all of your working.

SECTION 2

1. The four-engined Airbus A380 "super jumbo" is the world's largest passenger aircraft.

Information relating to one of its take-off runs appears below.

Mass of aircraft	522 tonnes (1 tonne = 1000 kg)
Maximum thrust **per engine**	348 kN
Take-off speed, V_R	80 ms^{-1}
Decision speed for rejected take-off, V_1	75 ms^{-1}
Emergency braking deceleration	3·5 ms^{-2}

(a) The pilot accelerated the plane along a level runway using 75% of the maximum thrust available. Show that the initial acceleration of the A380 along a level runway was 2 ms^{-2}. **4**

(b) Although the thrust was kept constant the average acceleration for the whole take-off run was only 1·6 ms^{-2}. State one reason for this. **1**

(continued)

(c) Calculate the time for the plane to reach its take-off speed from rest if it accelerated at $1{\cdot}6\ \mathrm{ms}^{-2}$. **3**

(d) Should a serious fault occur, the pilot can abandon take-off and select emergency brakes in order to bring the plane to rest quickly. This can be done at any speed up to and including V_1. Using the acceleration of $1{\cdot}6\ \mathrm{ms}^{-2}$ and any other data provided, calculate the minimum length of runway required for a safe rejected takeoff. **7**

Total marks 15

2. The velocity/time graph below shows data from an experiment in which a pupil dropped a 60 g ball which bounced off the ground.

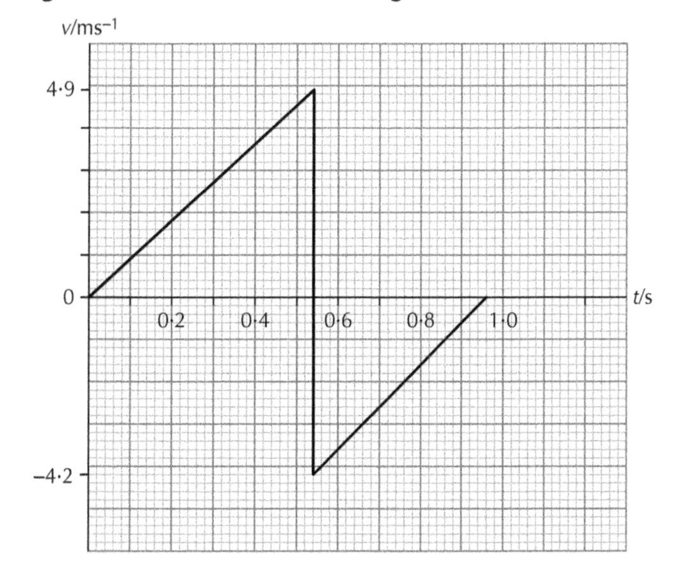

(continued)

(a) Calculate the change in momentum of the ball as a result of the bounce. **3**

Space for working and answer

(b) The ball was in contact with the ground for 2 ms. Show that the magnitude of the resultant force on the ball was 273 N. **3**

Space for working and answer

(c) Show by calculation that the ball lost approximately 26·4% of its kinetic energy as a result of the bounce. (Ignore energy losses due to any other causes.) **8**

Space for working and answer

(d) The student performed the experiment 10 times and obtained the results in the table below.

% E_k lost in bounce	26·5	24·0	29·2	28·8	24·0
	24·2	25·1	24·4	28·9	27·9

Calculate (i) An appropriate value for the percentage energy lost in the bounce. **1**

(continued)

(ii) The absolute uncertainty in the value calculated in part (i). 3

Space for working and answer

(e) A pupil looked at the answer to part (b) of the above question and said, "The resultant force on the ball was 273 N. That means the ground exerted an upwards force of 273 N on the ball and therefore the ball exerted an equal and opposite downwards force on the ground." 3

Using your knowledge of physics, comment on the pupil's statement.

Total marks 21

3. During a project on energy efficiency, a student makes the following suggestion:

"The effectiveness of an energy storage device could be assessed by finding the height to which it would rise if all its stored energy could be used to move it upwards."

Calculate the equivalent height for:

(a) (i) A 1·0 microfarad capacitor of mass 20 g charged to 5 kV. 3

(ii) A 20 kg car battery which can light a 100 W car headlamp for 6 hours. 3

(continued)

(b) A student fires a tennis ball vertically upwards during an experiment to find the effects of air resistance on the time of flight of the ball. The student suggests that in a vacuum, the time to travel up and the time to travel down would be equal. In air, this would still be true but the total flight time would be longer.

Use your knowledge of physics to comment on the student's claim. **3**

Total marks 9

4. The diagram below was taken from the calcium absorption spectrum of spiral galaxy 1357.

irradiance

396·2

wavelength/nm

(a) The wavelength of the calcium line measured on Earth is 393·4 nm.

(i) Calculate the velocity of the galaxy. **4**

Space for working and answer

(continued)

(ii) State and explain whether the galaxy is approaching or receding from Earth. **2**

(b) Put your answer to (a) (i) into the table below and use all the data to plot a graph of velocity against distance from Earth to the various galaxies using the graph paper provided at the end of the book. **3**

Galaxy number	***Distance from Earth/km × 10^{20}***	***Velocity km s^{-1}***
1357	8·6	
00139	17·6	4050
5548	20·7	5270
1540	26·2	5990

(c) Explain how your graph can be used to make an estimate of the Hubble Constant, H_0. **1**

(d) Use the graph to find a value for H_0. **3**

Total marks 13

5. A student sets up the circuit below to measure the value of a capacitor, C.

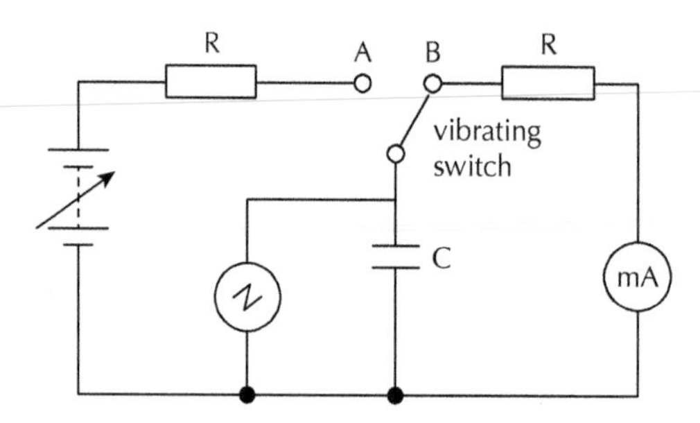

The capacitor is repeatedly charged and discharged by a vibrating switch. The oscilloscope can be used to determine the voltage across the capacitor and the frequency of the switching. The time, T, taken for the capacitor to charge can be found from $T = 1/f$, where f is the frequency of the vibrating switch.

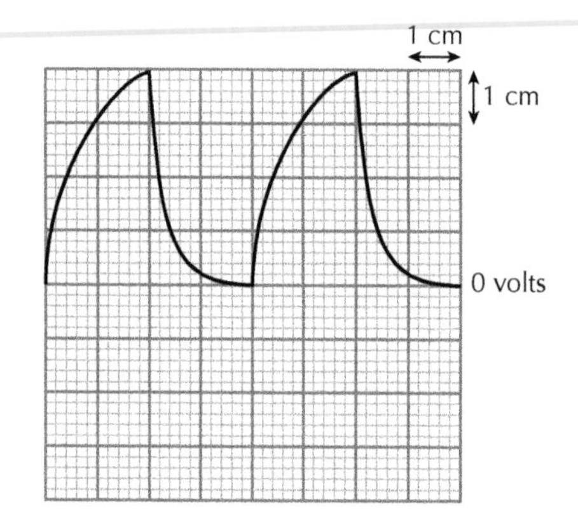

The timebase of the oscilloscope is set at 5 ms per cm. The Y gain is set at 2V per cm.

(a) Find the voltage across the capacitor when it is fully charged. **1**

Space for working and answer

(b) Show that the frequency of the vibrating switch is 50 Hz. **3**

Space for working and answer

(continued)

(c) The milliammeter reads 40 mA. Calculate the value of capacitance, C. **3**

Space for working and answer

Total marks 7

6. A specialist glass manufacturer measures the refractive index (n) of the glass using light of different frequencies. The numerical values shown in the table are for red light and violet light in air.

$f \times 10^{14}$ *Hz*	*n*
4·3	1·94
7·0	2·06

(a) Using the values in the table and any other data you require from the datasheet find

(i) the wavelength of red light in air. **3**

Space for working and answer

(ii) the wavelength of red light in the glass. **3**

Space for working and answer

MARKS | DO NOT WRITE IN THIS MARGIN

(continued)

(b) Narrow rays of red and violet light pass into two isosceles prisms made of the glass. Parts of the paths of the rays are shown below.

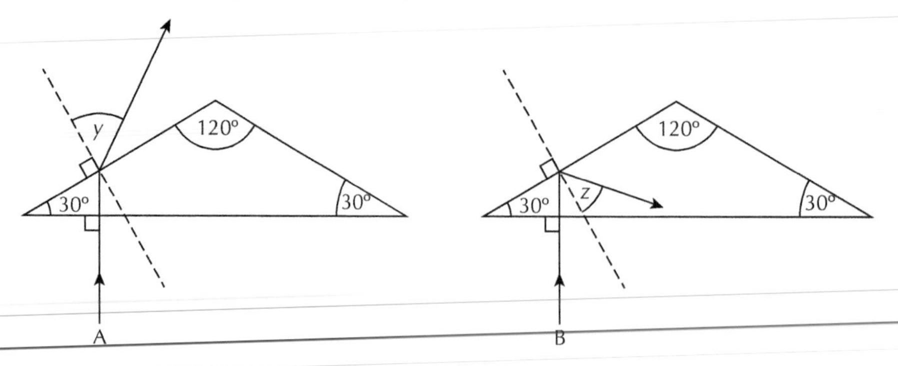

(i) Show by calculation that beam A is red light and beam B is violet light. **4**

Space for working and answer

(ii) Find the size of angles y and z. **3**

Space for working and answer

Total marks 13

7.

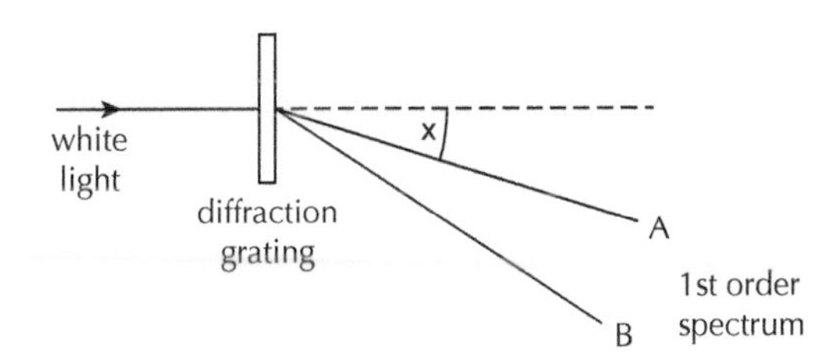

A narrow beam of white light is passed through a diffraction grating with $6{\cdot}0 \times 10^5$ lines per metre spacing.

(a) The wavelength of the light at A is 410 nm. Calculate the value of angle x. **4**

Space for working and answer

(b) If the grating is removed and a glass prism used to produce a spectrum between A and B, state one difference between this spectrum and the one produced by the grating. **1**

Total marks 5

8. Cosmic ray protons entering the upper atmosphere collide with nuclei to produce showers of pions (π) which decay to form muons (μ), electrons (e), neutrinos (ν) and anti-neutrinos ($\bar{\nu}$). Part of the process is illustrated in the diagram below.

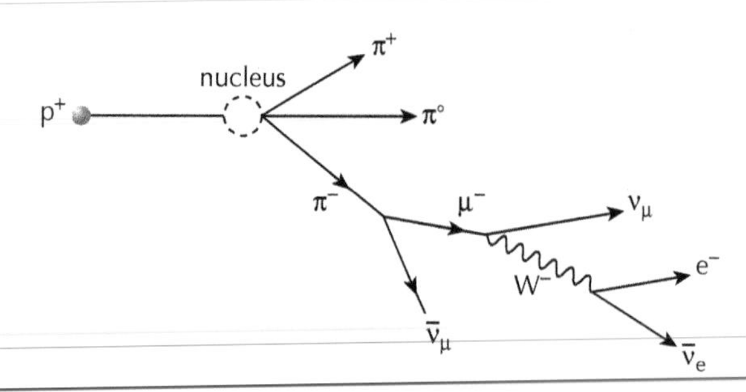

(a) Protons are composed of three quarks; pions are composed of a quark, antiquark pair.

	up quark ***u***	***down quark*** ***d***
charge	$+\frac{2}{3}$	$-\frac{1}{3}$

Using the information in the table determine the composition of

(i) a proton p^+

(ii) a pion, π^- **2**

(b) The muons are created at an altitude of approximately 15 km. They travel at a constant speed of 99·95% the speed of light (0·9995c) in the Earth's atmosphere. Laboratory measurements of the lifetime of the muons before decaying provides a mean value of $2{\cdot}2 \times 10^{-6}$ s.

(i) Using a simple classical physics formula, show that the muons should travel less than 700 m before decay and therefore not reach the surface of the Earth. **3**

Space for working and answer

(continued)

(ii) Using an appropriate formula from Special Relativity, calculate the mean lifetime of the muons relative to Earth and show that the muons will actually reach the Earth before they decay. **6**

Space for working and answer

Total marks 11

9. The table below gives the values of some energy levels in the hydrogen atom.

Level	*Energy*/10^{-20} J
5	−8·71
4	−13·6
3	−24·2
2	−54·2

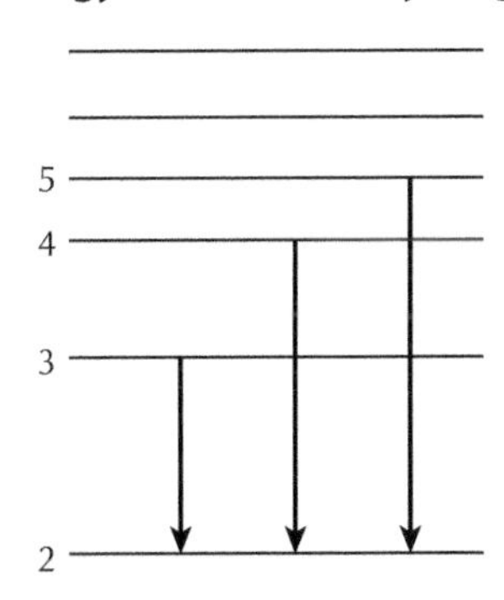

(a) Explain why the energy values are negative in the table. **2**

(b) (i) State which of the three transitions shown in the energy level diagram will produce a line in the emission spectrum with the longest wavelength. **1**

(continued)

(ii) Calculate the wavelength of the line. **4**

Space for working and answer

Total marks 7

10. A student is investigating the motion of a pendulum. The pendulum is pulled sideways and released. The pendulum's maximum vertical displacement (h) and its speed (v) at its lowest points are measured.

The results are shown in the table:

h/m	v/ms⁻¹
0·02	0·63
0·04	0·90
0·06	1·1
0·08	1·3
0·10	1·4

(a) By considering conservation of energy, show that for the pendulum, $v^2 = 2gh$. **2**

(b) Draw an appropriate graph and use the gradient to find a value for g, the acceleration due to gravity using the graph paper provided at the end of the book. **4**

(continued)

(c) A different student repeats the experiment measuring h as shown below:

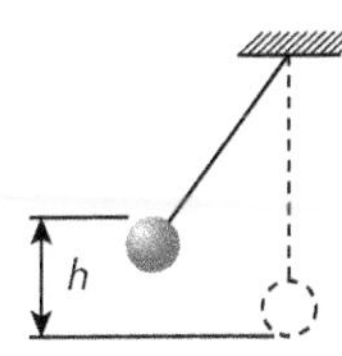

(i) What is the name given to the type of uncertainty that results from this measurement of h? **1**

(ii) Explain whether this uncertainty would affect the value of g. **2**

Total marks 9

[END OF QUESTION PAPER]

Practice Exam C

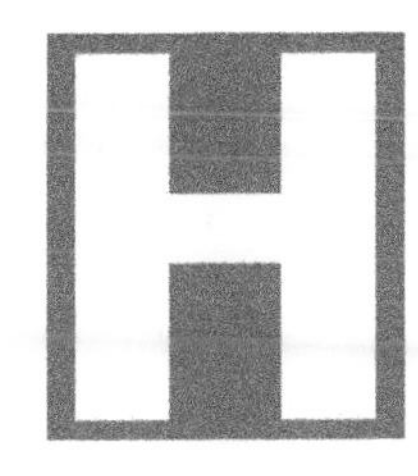

Higher Physics

Practice Papers for SQA Exams

Physics Section 1

Fill in these boxes:

Name of centre

Town

Forename(s)

Surname

Try to answer all of the questions in the time allowed.

Total marks – 130

Section 1 – 20 marks

Section 2 – 110 marks

Read all questions carefully before attempting.

You have $2\frac{1}{2}$ hours to complete this paper.

Write your answers in the spaces provided, including all of your working.

SECTION 1 ANSWER GRID

Mark the correct answer as shown ✓

	A	B	C	D	E
1	○	○	○	○	○
2	○	○	○	○	○
3	○	○	○	○	○
4	○	○	○	○	○
5	○	○	○	○	○
6	○	○	○	○	○
7	○	○	○	○	○
8	○	○	○	○	○
9	○	○	○	○	○
10	○	○	○	○	○
11	○	○	○	○	○
12	○	○	○	○	○
13	○	○	○	○	○
14	○	○	○	○	○
15	○	○	○	○	○
16	○	○	○	○	○
17	○	○	○	○	○
18	○	○	○	○	○
19	○	○	○	○	○
20	○	○	○	○	○

SECTION 1

1. A hot air balloon is rising with a constant velocity of 1·5 ms^{-1} when a sandbag accidentally breaks loose and falls to the ground. The sandbag is observed to hit the ground 3 seconds later.

The speed of the sandbag as it reaches the ground is

A 27·9 ms^{-1}

B 28·5 ms^{-1}

C 29·4 ms^{-1}

D 30·0 ms^{-1}

E 30·9 ms^{-1}.

2. In which two graphs below are the displacements identical?

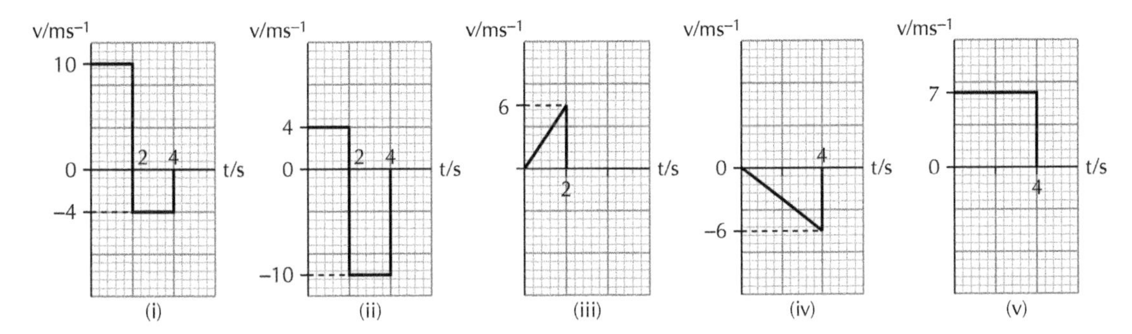

A (i) and (ii)

B (i) and (iii)

C (ii) and (iv)

D (ii) and (v)

E (iii) and (v).

3. A student is carrying out an investigation on the impulse imparted to a football when it is struck by a football boot. The ball is struck from rest and immediately passes through a lightgate which allows its final speed to be calculated.

Information gathered from the experiment is shown below.

Mass of football (0·75 ± 0·02) kg;
Diameter of football (219·6 ± 0·5) mm;
Time taken for football to pass through lightgate (0·022 ± 0·001) s;
Time of contact between the football boot and the football (0·031 ± 0·001) s.

The average force exerted by the boot on the ball and the uncertainty in this force is

	Average force/N	***Uncertainty/N***
A	70·2	±7·6
B	241·5	±11
C	241·5	±62
D	340·2	±37
E	340·2	±87

4.

Philae Lander

The Philae craft which landed on comet 67P in November 2014 has a mass of 100 kg. The craft was held on the surface of the comet 2 km from its centre by a gravitational force of $1{\cdot}67 \times 10^{-2}$ N. The mass of the comet is

A 1×10^{13} kg

B 5×10^{9} kg

C 1×10^{7} kg

D 5×10^{6} kg

E 1×10^{4} kg.

5. A spaceship is travelling at a speed of $2{\cdot}2 \times 10^{8}$ ms^{-1}. The spaceship emits a pulse of light that lasts for $5{\cdot}8 \times 10^{-5}$ s according to a stationary observer on Earth.

The duration of the pulse according to an astronaut on board the spaceship is

A $2{\cdot}7 \times 10^{-5}$ s

B $3{\cdot}0 \times 10^{-5}$ s

C $3{\cdot}9 \times 10^{-5}$ s

D $8{\cdot}5 \times 10^{-5}$ s

E 13×10^{-5} s.

6. A siren emits sound at a constant frequency of 12 kHz. A student attaches the siren to a horizontal turntable which moves the siren in a circle at a constant speed of 20 ms^{-1}. The speed of sound in air is 340 ms^{-1}. As the siren moves towards and away from the observer, the range of frequencies detected is

A 11294 to 12706 Hz

B 11294 to 12750 Hz

C 11333 to 12706 Hz

D 11333 to 12750 Hz

E 11980 to 12020 Hz.

7. When a clean metal plate is irradiated with a light source, photoelectrons are liberated from the surface of the metal.

The irradiance of the light source is now increased. This will

A increase the average kinetic energy of the photoelectrons but not increase the number of photoelectrons liberated.

B increase the average kinetic energy of the photoelectrons but decrease the number of photoelectrons liberated.

C increase the number and average kinetic energy of the photoelectrons.

D have no effect on either the number of photoelectrons liberated or the average kinetic energy of the photoelectrons.

E increase the number of photoelectrons but not increase the average kinetic energy of the photoelectrons.

8. Radiation of different frequencies hits a clean metal plate in a vacuum tube causing the emission of photoelectrons which are captured by another metal plate connected in a circuit as shown.

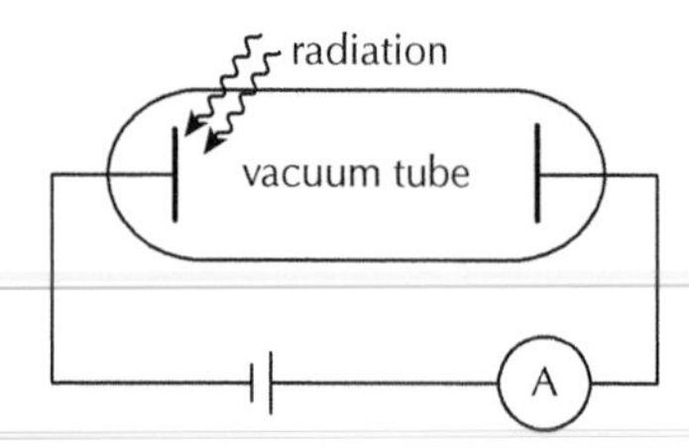

Which graph illustrates the correct relationship between current I and frequency f?

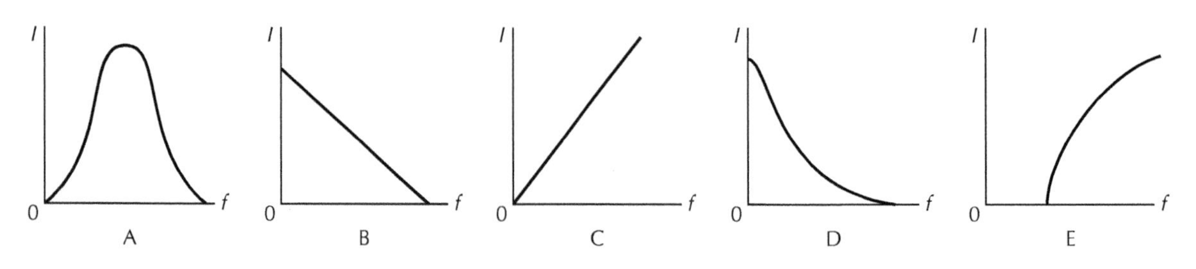

9. A ray of white light passes from air into a crown glass prism.

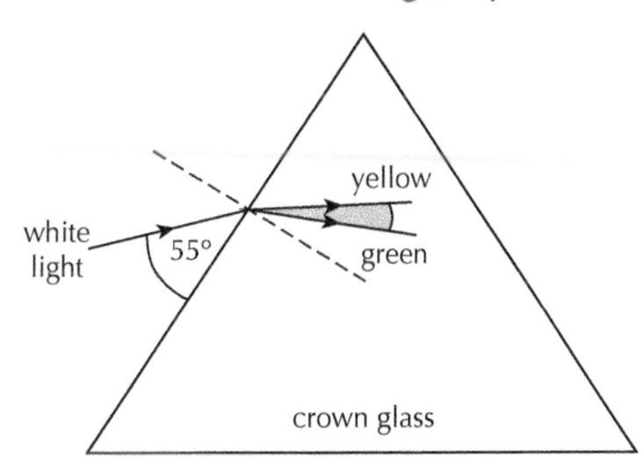

The refractive index of yellow light in crown glass is 1·511.

The refractive index of green light in crown glass is 1·515.

The angular separation of the rays of yellow and green light as they enter the glass prism is

A 0·004°

B 0·06°

C 0·097°

D 0·26°

E 0·61°.

10. The dwarf planet Pluto travels around the sun in an elliptical orbit. When it is nearest to the Sun at a distance of $4{\cdot}44 \times 10^{12}$ m, the irradiance on the surface of Pluto is 1·55 Wm^{-2}. When Pluto is furthest from the sun at $7{\cdot}32 \times 10^{12}$ m the irradiance on the surface of Pluto will be

A 0·57 Wm^{-2}

B 0·94 Wm^{-2}

C 1·55 Wm^{-2}

D 2·04 Wm^{-2}

E 4·21 Wm^{-2}.

11. A student makes three statements in a notebook about the Standard Model:

(i) Force mediating particles are bosons.

(ii) Fermions consist of 3 types of quark and leptons.

(iii) Mesons are made of three quarks.

Which of these statements is/are true?

A i only

B ii only

C iii only

D i and ii only

E ii and iii only

12. The diagrams show rays of light inside three transparent blocks with the refractive indices shown. The rays meet the face of each block at an angle of incidence of 42 degrees. Total internal reflection will occur in

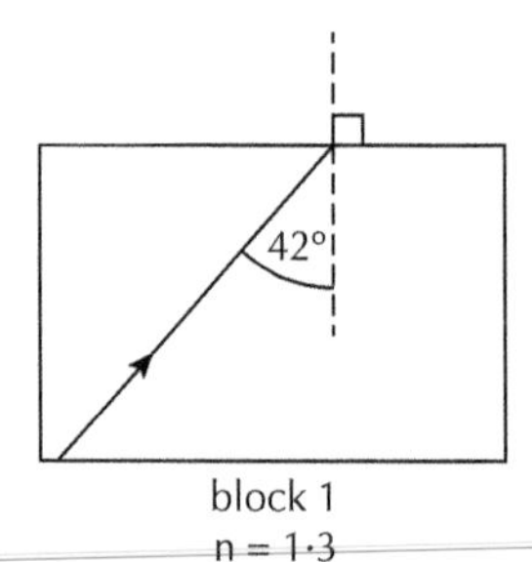

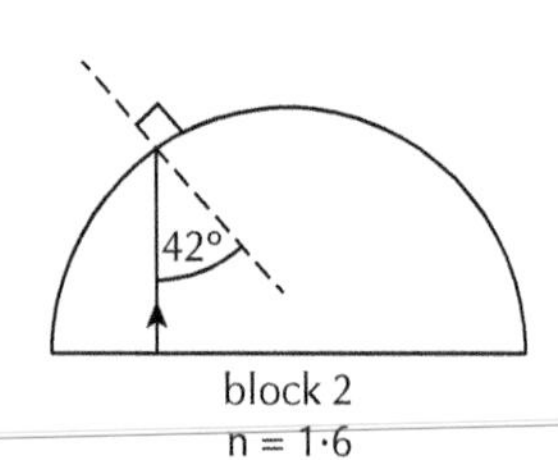

42°

block 3
n = 2·4

A block 1 only

B block 3 only

C blocks 1 and 2

D blocks 2 and 3

E all of the blocks.

13. A student draws diagrams in a notebook relating to the direction of field lines around point charges.

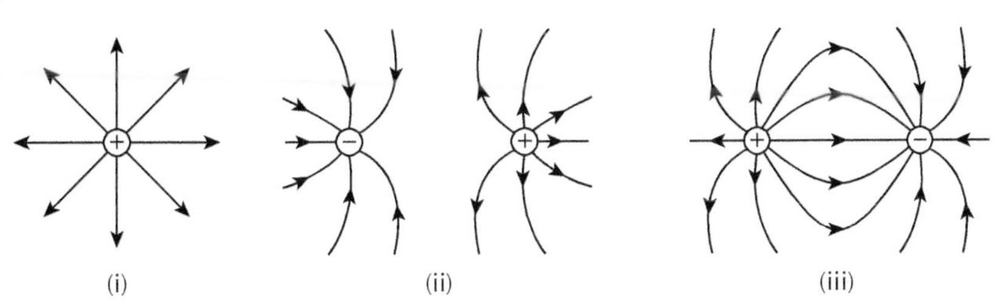

Which of the diagrams is/are correct?

A i only

B ii only

C iii only

D i and ii only

E i and iii only

14. An ohmmeter is connected to various points in the circuit below. Which row in the table shows a correct reading of resistance (to the nearest ohm) between each of the connection points?

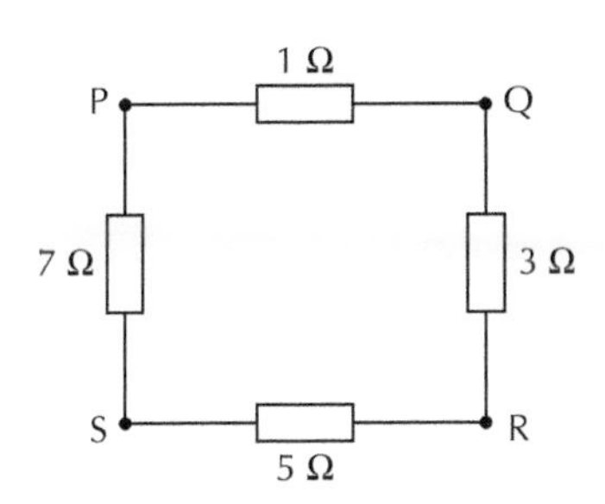

	Connection points	Ohmmeter reading
A	S and R	4 ohms
B	P and R	3 ohms
C	P and S	3 ohms
D	Q and S	8 ohms
E	P and Q	16 ohms

15. An electron enters a magnetic field as shown.

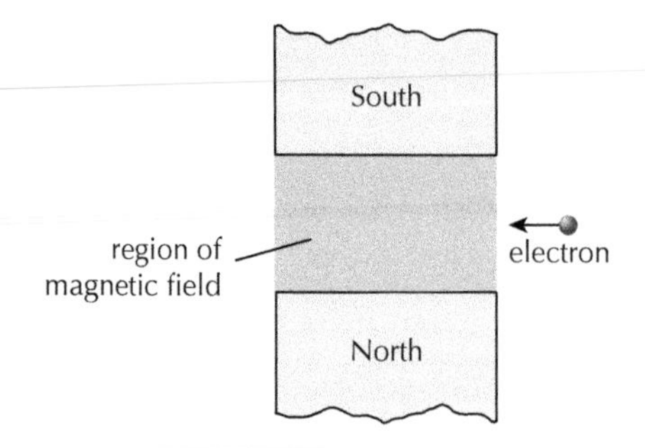

The electron will move

A towards the top of the page

B towards the bottom of the page

C into the page

D out of the page

E straight through from right to left.

16.

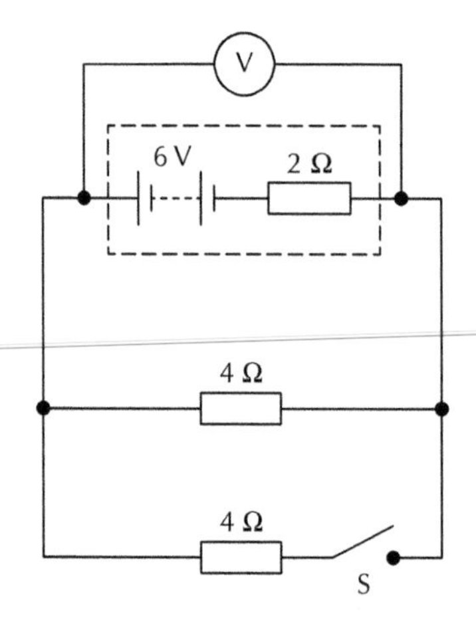

A battery has an emf of 6V and internal resistance 2 ohms. It is connected to two 4 ohm resistors, as shown. Switch, S is initially open. When S is closed the voltmeter reading

A remains at 6 V

B falls from 6 V to 4 V

C falls from 6 V to 3 V

D falls from 4 V to 3 V

E rises from 2 V to 3 V.

17. Two batteries and two lamps are connected together as shown.

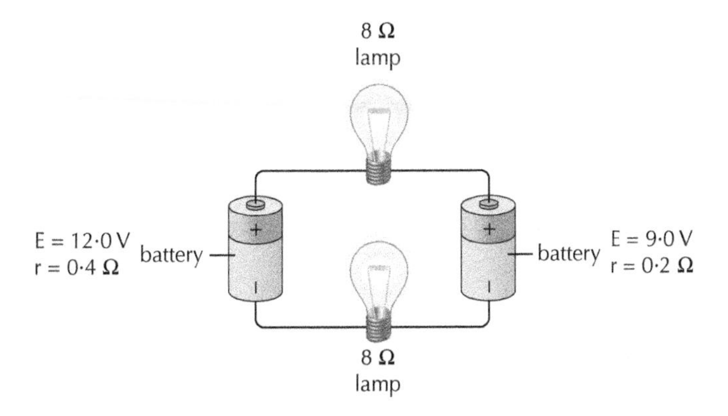

The current in this circuit when the lamps are operating normally is

A 0 mA

B 180·7 mA

C 185·2 mA

D 1024 mA

E 1049 mA.

18. An oscilloscope is used to display an a.c. waveform.

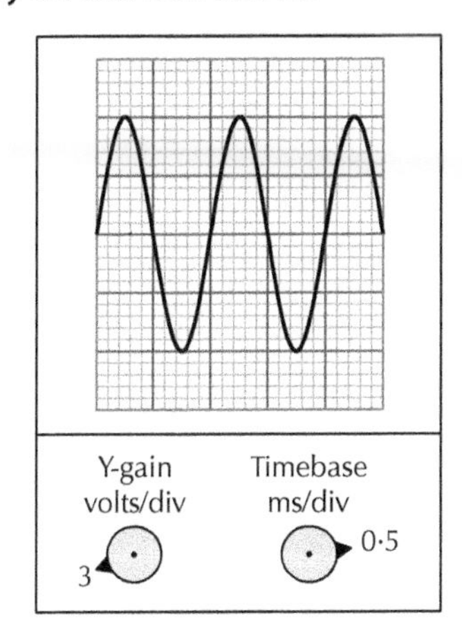

The rms voltage of the supply displayed on the oscilloscope is

A 4·2 V

B 6·0 V

C 8·5 V

D 12 V

E 17 V.

19. A circuit is set up by a student as follows:

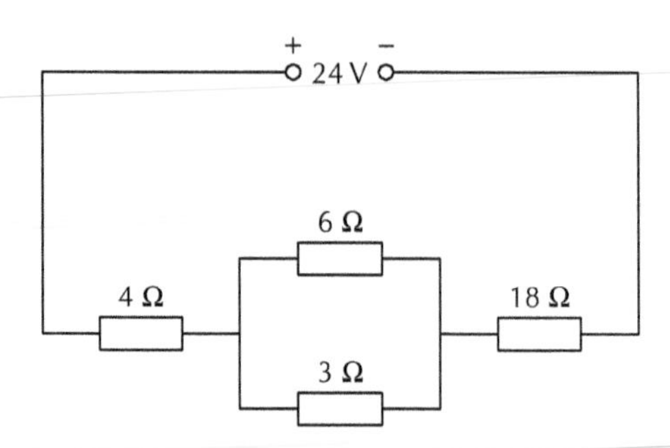

Adding an additional 2 Ω resistor in parallel with the 3 Ω resistor will

A decrease the p.d. across the 4 Ω resistor

B increase the p.d. across the parallel arrangement of resistors

C increase the circuit resistance

D decrease the circuit current drawn from the battery

E increase the circuit current drawn from the battery.

20.

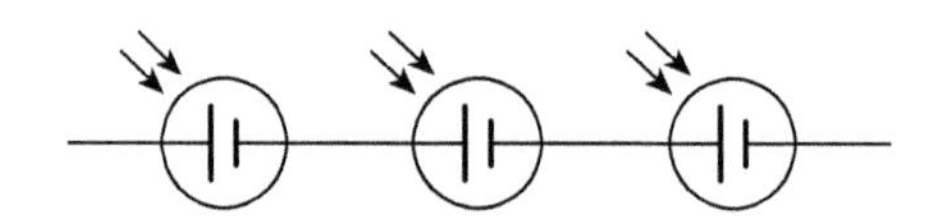

A battery consists of three p-n junction solar cells connected in series. It is illuminated by light of frequency 6×10^{14} Hz. The maximum output voltage of the battery will be

A 0·83 V

B 2·5 V

C 6·4 V

D 7·5 V

E 19 V.

Higher Physics

Practice Papers for SQA Exams

Physics Section 2

Fill in these boxes:

Name of centre

Town

Forename(s)

Surname

Try to answer all of the questions in the time allowed.

Total marks – 130

Section 1 – 20 marks

Section 2 – 110 marks

Read all questions carefully before attempting.

You have $2\frac{1}{2}$ hours to complete this paper.

Write your answers in the spaces provided, including all of your working.

MARKS | DO NOT WRITE IN THIS MARGIN

SECTION 2

1. A golfer strikes a golf ball with a velocity of 7·2 ms^{-1} at an angle of 50° to the horizontal in order to get it out of a bunker. The ball reaches its maximum height at point B which is 1·02 m above the level green next to the bunker. The ball first strikes the green at point C.

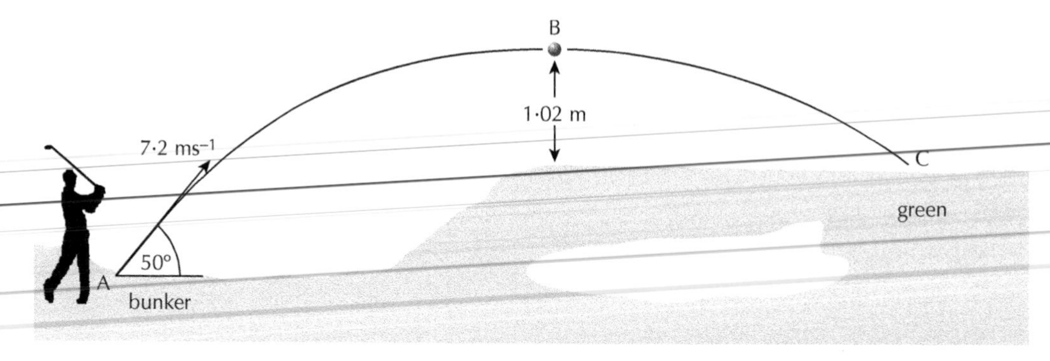

(a) (i) Calculate the horizontal component of velocity when the ball is struck. **1**

Space for working and answer

(ii) Calculate the vertical component of velocity when the ball is struck. **1**

Space for working and answer

(continued)

(b) Calculate the time taken for the ball to travel from where it is struck (point A) until it first touches the ground again (point C). **7**

Air resistance can be ignored.

Space for working and answer

(c) Calculate the horizontal distance travelled by the ball between point A and point C. **3**

Space for working and answer

Total marks 12

2.

An Edinburgh tram has a mass of 56,000 kg. It can carry up to 250 passengers. (The average mass of a passenger is 80 kg. You may ignore any frictional forces in this question.)

(a) A tram fully loaded with passengers accelerates from rest to 12 ms^{-1} in 10 s on a level track. Calculate the resultant force required to achieve this acceleration. **5**

Space for working and answer

(continued)

(b) Calculate the average power developed in part (a). **3**

Space for working and answer

(c) Later in the journey, with all of the passengers still on board, the tram again accelerates from rest to 12 ms^{-1} in 10 seconds, this time up a slope at an angle of 4·5°. Calculate the **additional** force required to achieve this acceleration. **3**

Space for working and answer

(d) The trams are fitted with electric motors which operate using a 750 V DC supply. Calculate the average current drawn by the tram in part (a) **3**

Space for working and answer

Total marks 14

3.

The new Queensferry Crossing over the Forth is a cable-stayed bridge. In this design each deck section is supported by a pair of cables which share the section load equally.

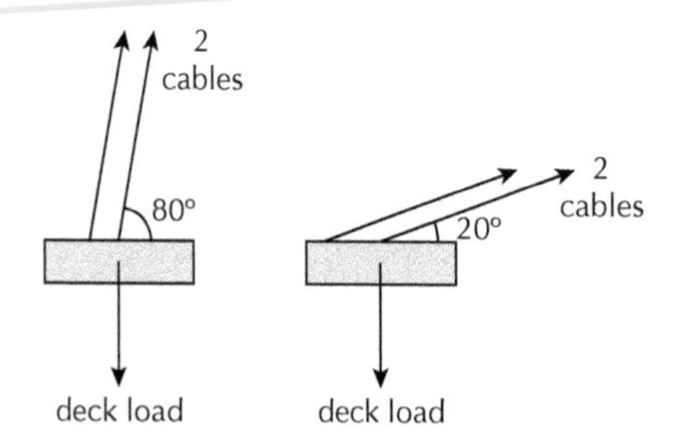

The pair of cables nearest the towers are at an angle of 80° to the deck. Cables furthest away are at an angle of 20° to the deck.

The total vertical loading on each deck section due to its own weight, the weight of traffic and wind effects is 1×10^7 N. Find the range of tensions which would be found in the cables of the bridge. **5**

Space for working and answer

Total marks 5

4. (a) Newspaper articles sometimes refer to astronauts on board the International Space Station as being *weightless*.

Use your knowledge of Physics to comment on this statement. **3**

Space for working and answer

(b) Whilst on a spacewalk an astronaut pushes against the side of the International Space Station. The velocity of both the astronaut and the Space Station change as a result.

Explain this phenomenon. **2**

Space for working and answer

(c) The International Space Station can change its direction by firing jets of nitrogen gas for short periods of time.

Explain how this changes the direction of the Space Station. **2**

Space for working and answer

Total marks 7

5. While a physicist is celebrating his 30th birthday on Earth, his twin brother who is an astronaut flies past the Earth in a spacecraft at a constant speed of $2{\cdot}6 \times 10^8$ ms^{-1} and continues on his way into deep space. Ten years later, on his 40th birthday, the physicist comments that his twin will only be 35 years old.

(a) Setting out clearly any numerical calculations, show how the physicist calculated the age of his astronaut brother. **4**

Space for working and answer

(b) During his journey into deep space at $2{\cdot}6 \times 10^8$ ms^{-1}, the astronaut performs two experiments to measure the speed of light. The first experiment is conducted using apparatus entirely inside his spacecraft; the second makes use of a laser beam which is reflected off a nearby planet. Explain how the two readings should compare. **2**

Space for working and answer

(continued)

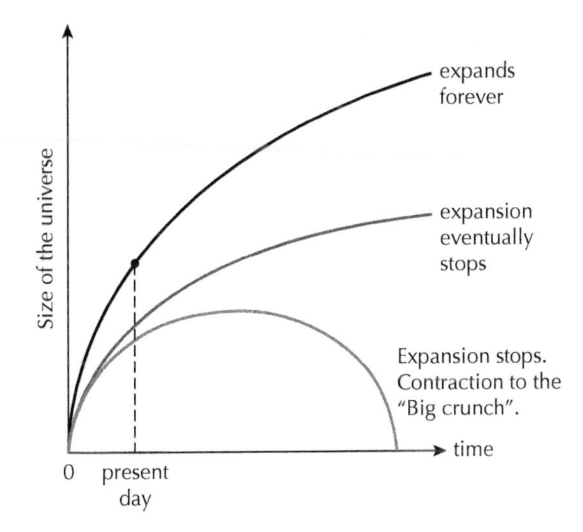

(c) During a public lecture on possible futures for our universe, a cosmologist shows the above slide to the audience.

Use your knowledge of physics to comment on the three scenarios shown in the slide. **3**

Total marks 9

MARKS | DO NOT WRITE IN THIS MARGIN

6. A poster in a classroom shows some members of the 'Particle Zoo'.

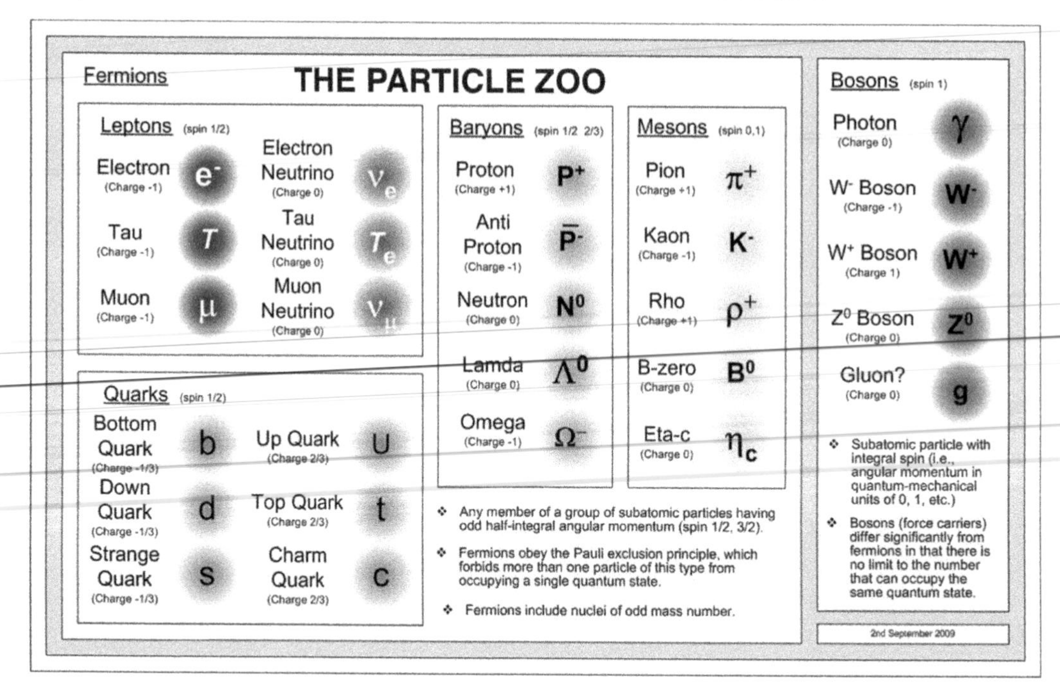

(a) One member of the Particle Zoo, the K^- Kaon is known as an antiparticle. State what is meant by the term *antiparticle*? **1**

(b) The K^- Kaon is also classed as a *meson*. Describe the composition of a meson. **1**

(c) What is the quark configuration of a Pi (π^+) meson if it is composed of only *'up'* and *'down'* quarks? **2**

(d) What is the name of the force mediating particle associated with the Electromagnetic Force? **1**

(continued)

(e) Over what range does the Gravitational Force extend? **1**

(f) What type of decay was the first evidence for the existence of the neutrino? **1**

Space for working and answer

Total marks 7

7. In the diagram below, a subatomic particle of mass $6{\cdot}65 \times 10^{-27}$ kg and charge $3{\cdot}20 \times 10^{-19}$ C travelling at $2{\cdot}53 \times 10^{6}$ ms^{-1} in a vacuum chamber is shown passing through a small hole in plate A, travelling horizontally across the electric field between the plates and exiting through plate B to reach a detector.

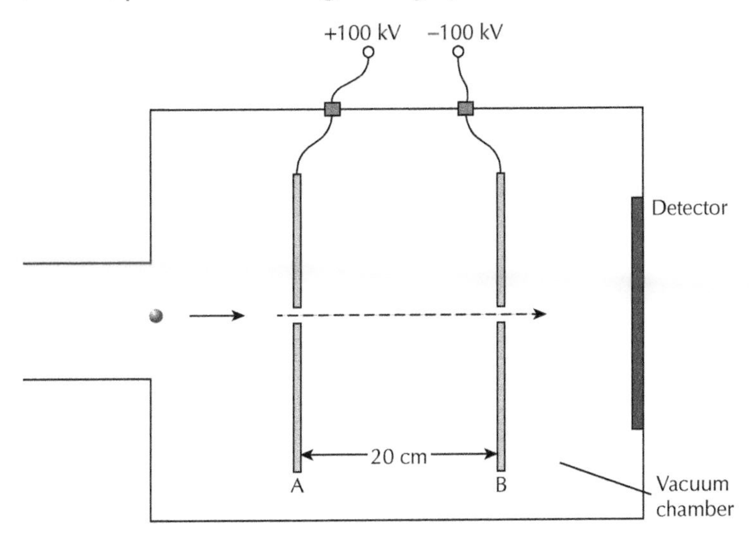

(a) What is the potential difference across the plates? **1**

Space for working and answer

(continued)

(b) The electric field between plates A and B is uniform.

(i) Explain what is meant by the term 'uniform'. **1**

(ii) Make a sketch of the plates and show the electric field lines between the plates. **2**

(c) By considering the kinetic energy of the particle as it enters the field and the work done by the field, show that the speed of the particle will have doubled by the time it reaches plate B. **4**

Space for working and answer

MARKS | DO NOT WRITE IN THIS MARGIN

(continued)

(d) Calculate the force acting on the particle while it travels between the plates. **3**

Space for working and answer

Total marks 11

8. The following statement describes a reaction that takes place in a nuclear fission reactor.

$$^{1}_{0}n + ^{235}_{92}U \rightarrow ^{137}_{55}Cs + ^{95}_{37}Rb + 4^{1}_{0}n$$

Data relating to the particles involved is shown in the table.

Particle	***Mass (kg)***
$^{1}_{0}n$	$1{\cdot}673 \times 10^{-27}$
$^{235}_{92}U$	$390{\cdot}219 \times 10^{-27}$
$^{137}_{55}Cs$	$227{\cdot}291 \times 10^{-27}$
$^{95}_{37}Rb$	$157{\cdot}562 \times 10^{-27}$

(a) What is meant by *nuclear fission*? **1**

(b) Calculate the energy released in this reaction. **4**

Space for working and answer

MARKS

DO NOT WRITE IN THIS MARGIN

(continued)

(c) The power station generates 600 MW of power using this reaction. Calculate the average number of fission reactions occurring each second. **3**

Space for working and answer

(d) Another type of nuclear reaction is nuclear fusion.

Describe in detail **two** problems associated with the production of energy by nuclear fusion. **4**

Total marks 12

9.

The speed limit for space vehicles in a distant solar system is $2{\cdot}0 \times 10^7$ ms^{-1}. The driver of a rocket racer is in court accused of driving through a red light at a planetary crossroads. His defence is that although the light may have been at red ($\lambda = 650$ nm) his approach speed meant that he saw the light as amber ($\lambda = 600$ nm) due to blue shift. The judge decides to acquit him of deliberately travelling through a red light but fines him for speeding. Explain clearly, showing any numerical calculations, why the judge decides that the driver must have been speeding. **4**

Total marks 4

10. A microwave transmitter, an aluminium reflector, a metre rule, a microwave receiver and a voltmeter are used to investigate interference.

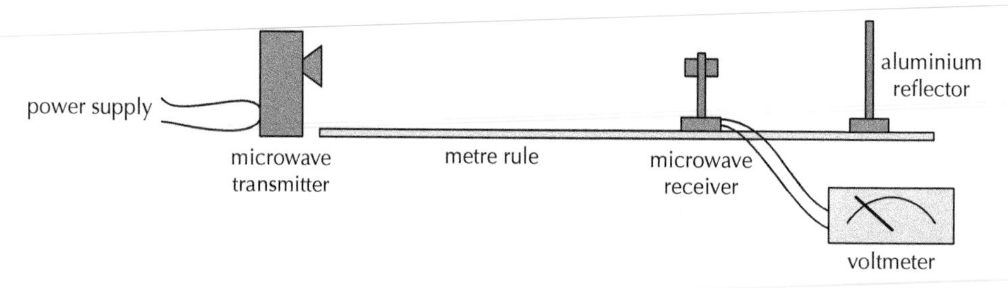

(a) Describe in detail how the apparatus produces an interference effect. **5**

The receiver is placed at a position of maximum interference and its position noted using the metre rule. The receiver is now moved closer to the transmitter passing through a further 9 maxima before stopping at the 10th maximum. The total distance moved by the receiver is 140 mm.

(b) (i) How many complete wavelengths are there over this 140 mm distance? **1**

Space for working and answer

(ii) Calculate the wavelength of the microwave radiation used in the experiment. **3**

Space for working and answer

Total marks 9

MARKS | DO NOT WRITE IN THIS MARGIN

11. The following diagram shows two 90° isosceles triangular glass prisms, A and B, with different refractive indices (n). Rays of monochromatic red light are directed at 90 degrees to the prisms as shown.

(a) Complete the diagrams by accurately drawing in the paths of the rays until they emerge into the air. Where the rays change direction, you should mark in clearly any angles. **9**

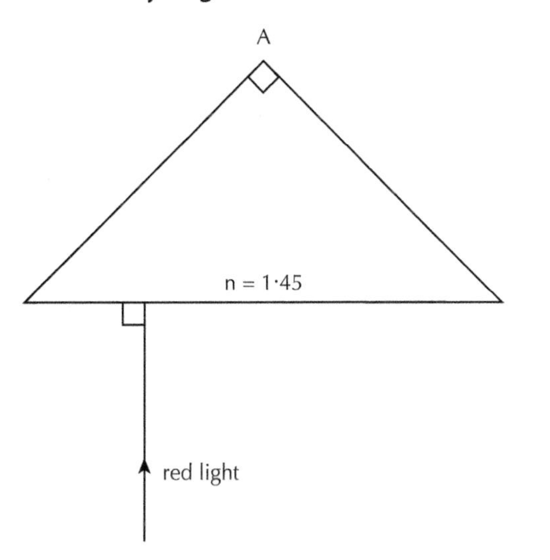

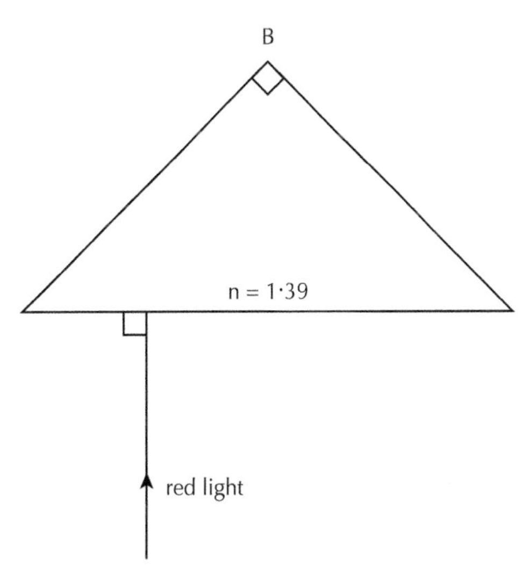

(b) The wavelength of the red light in air is 650 nm. Show that the wavelengths of the light in the two prisms differ by 19.4 nm. **4**

Space for working and answer

Total marks 13

MARKS | DO NOT WRITE IN THIS MARGIN

12. A circuit is set up to observe the effect of capacitance in an a.c. circuit.

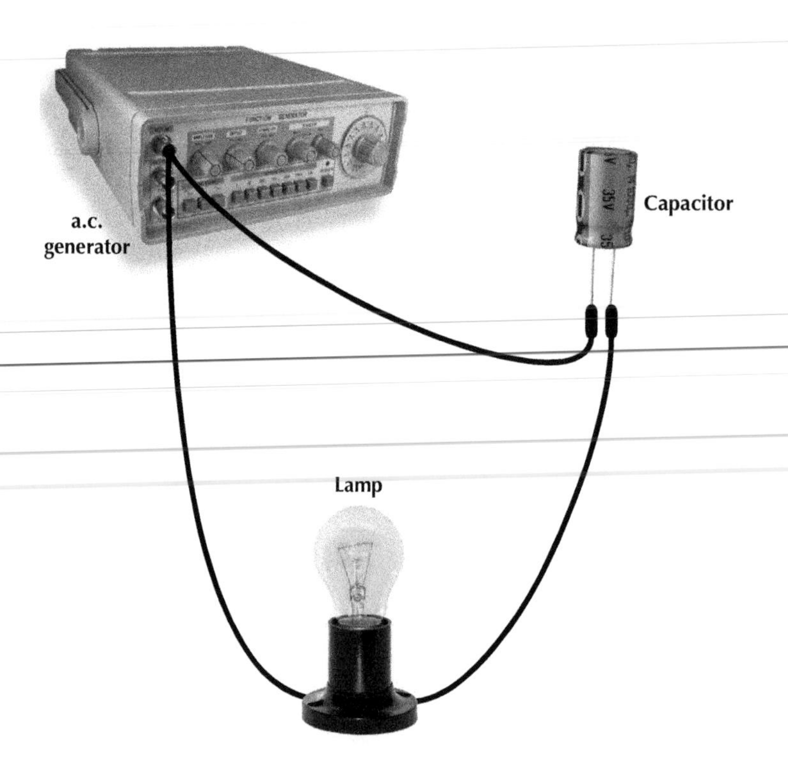

(a) Explain fully why the lamp lights continuously even though there is an insulator between the capacitor plates. **3**

(b) The frequency of the a.c. supply is increased. Describe and explain what happens to the brightness of the lamp. **2**

(continued)

(c) The capacitor is now replaced with a smaller value capacitor. Explain what would happen to the brightness of the lamp. **2**

Total marks 7

[END OF QUESTION PAPER]

Answers

Answers to Practice Exams

Practice Exam A

Section 1

Question	Response	Mark	Top Tips
1.	D	**1**	You have to use a = (v – u)/t for each 2 second time slot to see how the acceleration changes with time.
2.	E	**1**	The camera will always be in line with the launcher therefore any motion captured will be purely vertical. The ball slows down as it rises therefore each successive image is closer to the one following.
3.	B	**1**	There is a rapid deceleration (initially) due to travelling up a rough surface and then an acceleration down the smoother part of the ramp. The magnitude of the acceleration is smaller than the deceleration as the rough surface has more of an impact on the change in velocity than the smooth surface.
4.	A	**1**	To calculate the force applied you have to calculate the change of momentum per second, i.e., use F = m(v – u)/t.
5.	C	**1**	The acceleration of the entire train is calculated using $a = F_{unb}/m$. This acceleration (0·7 ms^{-2}) is then used with Tension = ma + $F_{friction}$ using only the mass and friction of the last carriage to calculate the tension in the coupling.
6.	A	**1**	Use $F = GMm/r^2$ and don't forget that the distance 'r' is the radius from the centres of each body. Also remember to square the distance.
7.	D	**1**	Time on clocks moving at close to the speed of light (*c*) is observed by a stationary observer to slow down. In addition, the length of an object gets shorter as the object approaches the speed of light.
8.	C	**1**	Take into account that the sound from the horn is approaching a stationary observer and use $f_0 = f_s\left[\frac{v}{(v-v_s)}\right]$.
9.	C	**1**	Mesons are intermediate mass particles which are made up of a quark-antiquark pair. Three quark combinations are called baryons. Mesons and baryons are both hadrons, which together with leptons, are fermions.
10.	B	**1**	Use $E_k = (hc)/\lambda - \Phi$ (where Φ = work function) then use $E_k = \frac{1}{2}mv^2$. **Care** – don't forget the square root when calculating 'v'.

11.	E	**1**	Start by using the 'right hand rule' which is associated with **negative** charges, that is, orientate the thumb and first two index fingers at 90° to each other. Now point the first finger in the magnetic field direction, the second finger in the direction of the velocity and the thumb will now indicate the direction of the force (on a negative charge). For a **positive** charge (as indicated in this question) simply reverse the direction of the thumb used with the 'right hand rule' above.
12.	E	**1**	The photoelectric effect is evidence for the particulate nature of light not line spectra. Wavelength inversely proportional to frequency ('c' is constant) therefore since threshold frequency f_0 is the minimum frequency required, threshold wavelength will be the maximum.
13.	C	**1**	Using $m\lambda = d \sin\theta$, $d = 1{\cdot}11 \times 10^{-6}$ m, $\theta_{max} = 90°$ and $\lambda = 500 \times 10^{-9}$ m gives $m = 2$ but there are 2 fringes on **each** side of the central fringe therefore 5 in total (including the central fringe).
14.	E	**1**	Amplitude is unaffected by a change in the medium through which a ray of light passes. Wavelength and wave speed both increase in passing from a higher refractive index to a lower refractive index material.
15.	C	**1**	Use path difference = $(m + ½)\lambda$ with $m = 2$ (3rd minimum) **N.B. not third 'order' minimum.**
16.	D	**1**	The emission of a photon comes about as the result of a downward transition between energy levels. Longest wavelength corresponds to smallest transition between energy levels.
17.	B	**1**	The amplitude is 2 div high therefore $2 \times$ Y-gain = 10V. The period is 2 div therefore $2 \times$ timebase = 200 ms and $f = 1/T = 1/200 \times 10^{-3} = 5$ Hz.
18.	A	**1**	The combined resistance should be found using $R_T = R/N$ where R = resistance of one of the **identical** resistors in parallel and N = number of resistors.
19.	B	**1**	Doping with Gallium creates positive holes which reduces the resistance of the silicon crystal and hence allows the crystal to conduct.
20.	A	**1**	

Practice Exam A

Section 2

Question	Expected Response	Mark	Max mark	Top Tips
1. (a)	$F_{unb} = ma$	1	3	
	$= 0{\cdot}40 \times 0{\cdot}80$	1		
	$= 0{\cdot}32$ N	1		
(b)	$F_{unb} = F_p - (F_{\parallel} + F_f)$	1	5	Remember that the unbalanced force = the difference between the forces acting in opposite directions.
	$0{\cdot}32 = 1{\cdot}98 - (F_{\parallel} + 0{\cdot}20)$	1		F_{unb} = unbalanced force
	$F_{\parallel} = 1{\cdot}46$ N			F_p = pulling force
	$F_{\parallel} = mg\sin\theta = 1{\cdot}46$	1		$F_{\parallel}$ = parallel component of the weight force
	$0{\cdot}40 \times 9{\cdot}8 \times \sin\theta = 1{\cdot}46$	1		F_f = friction
	$\theta = \sin^{-1}(1{\cdot}46/(0{\cdot}40 \times 9{\cdot}8) = 22°$	1		
(c)	Constant speed $\Rightarrow$ balanced forces		4	
	$F_p = (F_{\parallel} + F_f)$	1		
	$1{\cdot}98 = (F_{\parallel} + 0{\cdot}20)$	1		
	$F_{\parallel} = mg\sin\theta = 1{\cdot}78$			
	$0{\cdot}40 \times 9{\cdot}8 \times \sin\theta = 1{\cdot}78$	1		
	$\theta = \sin^{-1}(1{\cdot}78 / (0{\cdot}40 \times 9{\cdot}8))$			
	$\theta = 27°$	1		
2. (a)	Thrust = force + accelerating force		3	Remember that the total force is a combination of the weight force **and** the accelerating force.
	$= mg + ma$			
	$= m(g + a)$	1		
	$= (3850 + 650) \times (9{\cdot}8 + 2)$	1		
	$= 53100$ N	1		

(b)	$W = mg$ $= 650 \times 9{\cdot}8$ $= 6370$ N (triangle: T, 6370 N, 3678 N) $T = \sqrt{6370^2 + 3678^2}$ $= 7356$ N	 1 1 1 1	4	Basic vector addition problem involving horizontal and vertical components – remember the tension acts on the load.
(c) (i)	Thrust = weight force $= m \times g$ $= 3850 \times 9{\cdot}8$ $= 37{\cdot}7$ kN	 1 1 1	3	The load has been deposited therefore Newton's first law tells us that if the object is stationary then the forces must balance.
(ii)	$v^2 = u^2 + 2as$ $0 = 2^2 + (2 \times a \times 32)$ $a = 0{\cdot}06$ ms^{-2}	1 1	2	Standard equation of motion question – remember that "show that" question means working **must** be shown. In a "show that" type question always leave your final answer exactly as requested i.e., 0·06, **not** 0·063 or 0·0625.
(iii)	$\bar{u} = (u + v)/2$ OR $v = u + at$ (1) $= (2 + 0)/2$ $0 = 2 - (0{\cdot}0625t)$ (1) $= 1$ ms^{-1} $t = 32$ s (1) $\bar{u} = d/t$ $1 = 32/t$ $t = 32$ s	1 1 1	3	Remember that only the initial speed is 2 ms^{-1} and that $\bar{u} = d/t$ only applies to **average** speeds.
3. (a) (i)	Change in momentum = area under Ft graph $\Delta p = (5 \times 4) + ((3 \times 4)/2)$ $= 26{\cdot}0$ kg ms^{-1}	1 1 1	3	Take care when calculating areas – particularly triangles.

(ii)	$\Delta p = m(v - u)$	1	2	
	$26{\cdot}0 = 18{\cdot}0\ (v - u)$	1		
	Change in velocity $= 1{\cdot}44\ ms^{-1}$			
(b) (i)	Total momentum before = Total momentum after		3	Stones remain separate therefore RHS is $m_1v_1 + m_2v_2$. If they stuck together then RHS would be $(m_1 + m_2)v$.
	$m_1u_1 + m_2u_2 = m_1v_1 + m_2v_2$			
	$18{\cdot}0u_1 + 0 = (18{\cdot}0 \times 0{\cdot}40) + (20{\cdot}0 \times 0{\cdot}60)$	1		
	$18{\cdot}0u_1 = 19{\cdot}2$	1		
	$u_1 = 1{\cdot}07\ ms^{-1}$	1		
(ii)	Total E_k before $= \frac{1}{2} m_1u_1^2 = \frac{1}{2} \times 18{\cdot}0 \times 1{\cdot}07^2 = 10{\cdot}3$ J	1	3	Remember if E_k before > E_k after collision is inelastic, and if E_k before = E_k after collision is ELASTIC. All working **must** be shown.
	Total E_k after $= \frac{1}{2} m_1v_1^2 + ½\ m_2v_2^2 = \frac{1}{2} \times 18{\cdot}0 \times 0{\cdot}40^2 + \frac{1}{2} \times 20{\cdot}0 \times 0{\cdot}60^2 = 5{\cdot}04$ J	1		
	Since E_k before > E_k after, collision is INELASTIC.	1		
(c)	$F_{ave} = \Delta p/\Delta t = m(v - u)/t$	1	3	Remember average force = rate of change of momentum.
	$= 20{\cdot}0\ (0{\cdot}60 - 0)/10 \times 10^{-3}$	1		
	$= 1200$ N	1		
4.	Demonstrates no understanding 0 Limited understanding 1 Reasonable understanding 2 Good understanding 3 **1 mark:** The student has demonstrated a limited understanding of the physics involved. The student has made some statement(s) which is/are relevant to the situation, showing that at least a little of the physics within the problem is understood.	3	3	Open-ended question. Try and write down everything you know about the physics described in the question and, if possible, include a law of physics or equation.

	2 marks: The student has demonstrated a reasonable understanding of the physics involved. The student makes some statement(s) which is/are relevant to the situation, showing that the problem is understood. **3 marks:** The maximum available mark would be awarded to a student who has demonstrated a good understanding of the physics involved. The student shows a good comprehension of the physics of the situation and has provided a logically correct answer to the question posed. This type of response might include a statement of the principles involved, a relationship or an equation, and the application of these to respond to the problem. This does not mean the answer has to be what might be termed an 'excellent' answer or a 'complete' one.			
5. (a)	The star has a minimum wavelength when it is moving towards the Earth on one side of its orbit around the centre of mass and a maximum wavelength when moving away from the Earth on the other side of the orbit.	1 1	2	Remember stellar objects moving towards the Earth are blue-shifted and stellar objects moving away from the Earth are red-shifted.
(b) (i)	$v/c = (\lambda_{obs} - \lambda_{rest})/\lambda_{rest}$ $v/3{\cdot}0 \times 10^8 = ((660{\cdot}78 - 655{\cdot}96) \times 10^{-9}/660{\cdot}78 \times 10^{-9})$ $v = 2{\cdot}2 \times 10^6\ ms^{-1}$	1 1 1 1	4	$\Delta\lambda$ = the difference between the actual wavelength of the line and the wavelength of either line. Equation derived from a combination of $z = \frac{v}{c}$ and $z = \frac{\lambda_{observer} - \lambda_{rest}}{\lambda_{rest}}$ – both provided on formula sheet.

(ii)	$v/c = (\lambda_{obs} - \lambda_{rest})/\lambda_{rest}$ $2{\cdot}2 \times 10^6 / 3{\cdot}0 \times 10^8$ $= (\lambda - 434{\cdot}05 \times 10^{-9})/434{\cdot}05 \times 10^{-9}$ $= 437{\cdot}23$ nm	 1 1 1	3	Be careful not to get the measured wavelength and the actual wavelength mixed up.
6. (a)	An alternating p.d. is required to provide an oscillating electric field which will reverse the polarity across the gap each half revolution. This ensures that the particles are repelled by one dee and attracted towards the other dee.	1 1 1	3	
(b)	$qV = ½ mv^2$ $1{\cdot}6 \times 10^{-19} \times 28 \times 10^3 = \frac{1}{2} \times 1{\cdot}673 \times 10^{-27} \times v^2$ $v = 2{\cdot}3 \times 10^6$ ms^{-1}	1 1 1	3	Always remember to check the speed by comparing to the speed of light ($3{\cdot}0 \times 10^8$ ms^{-1}) because if your answer is greater than this value you've probably forgotten to take the square root.
(c)	v = d/t t is constant since π, m*, q and B are all constant. Thus as the speed, v increases, the distance travelled (C = 2πr) increases, i.e. the radius of curvature, r must increase resulting in an increasing spiral path.	 1 1 1	3	m* – this is of course assuming that the speed does not reach 10% of the speed of light otherwise relativistic effects must be considered.
(d)	**Any 2 from:** Electrons are much less massive than protons and would thus reach relativistic speeds far too quickly to be of any use. There is also a low probability that an electron could traverse the electron cloud surrounding the nucleus and therefore might not reach the nucleus. Finally, the electron is so light compared to the nucleus itself it would be like bouncing a ping-pong ball off a brick wall. Or other reasonable answer…	 1 1 1	3	

7. (a) (i)	Work function = hf_o	1	2	
	$4{\cdot}64 \times 10^{-19} = 6{\cdot}63 \times 10^{-34} \times f_o$	1		
	$f_o = 7 \times 10^{14}$ Hz			
(ii)	$E_k = (hc)/\lambda$ – work function	1	4	$hf = (hc)/\lambda$
	$= (6{\cdot}63 \times 10^{-34} \times 3{\cdot}0 \times 10^{8})/340 \times 10^{-9} - 4{\cdot}64 \times 10^{-19}$ $= 1{\cdot}21 \times 10^{-19}$ J	1		
	$E_k = ½\, mv^2$			
	$1{\cdot}21 \times 10^{-19} = ½ \times 9{\cdot}11 \times 10^{-31} \times v^2$	1		
	$v = 5{\cdot}15 \times 10^{5}$ ms^{-1}	1		
(b) (i)	Decreasing the irradiance will reduce the number of photons arriving and hence decrease the number of photoelectrons escaping.	1	3	
(ii)	It does not however have any influence upon the velocity of the liberated photoelectron,	1		
	as this is controlled by the wavelength of the photons and the work function of the metal neither of which has changed.	1		
8.	Demonstrates no understanding 0 Limited understanding 1 Reasonable understanding 2 Good understanding 3	3	3	Open-ended question. Try and write down everything you know about the physics described in the question and, if possible, include a law of physics or equation.
9. (a)	The first narrow slit is used to ensure that the light which reaches the double slits is coherent.	1	1	The light which emerges from each of the double slits has come from the same point (first slit) and will therefore be coherent. Coherent light is required before an interference pattern is observable.

(b) (i)	$D_{mean} = (1{\cdot}921 + 1{\cdot}927 + 1{\cdot}924 + 1{\cdot}922 + 1{\cdot}924)/5$ $= 1{\cdot}924$ m Random uncertainty $= (1{\cdot}927 - 1{\cdot}921)/5$ $= 1{\cdot}2 \times 10^{-3}$ m	 1 1	**2**	Do not be tempted to quote the mean distance to a greater number of significant figures than the data provided.
(ii)	Rearranging $m\lambda D = dw$ $\lambda = dw/mD$ $= (0{\cdot}126 \times 10^{-3} \times 15{\cdot}50 \times 10^{-3})/(2 \times 1{\cdot}924)$ $= 508$ nm % uncertainty in $D = 0{\cdot}062$ % (from (b) (i) above) % uncertainty in $w = (0{\cdot}05/15{\cdot}50) \times 100 = 0{\cdot}323\%$ % uncertainty in $d = (0{\cdot}001/0{\cdot}126) \times 100 = 0{\cdot}794\%$ Overall % uncertainty $= 0{\cdot}794\%$ (largest) Absolute uncertainty in $\lambda = 0{\cdot}794\%$ of 508 nm $= 4{\cdot}03$ nm Therefore, wavelength $\lambda = (508 \pm 4)$nm	 1 1 1 1 1 1	**6**	Although this formula is not provided on the formula sheet it is provided within the question and therefore can be used to set a question. This would be considered a 'problem solving' type question. Overall percentage uncertainty used in Higher Physics is always the largest of the individual uncertainties.
(iii)	Green	1	**1**	Use the date sheet to find the colour of the wavelength closest to the value in (b) (ii).
(c) (i)	w would increase.	1	**1**	Rearranging the equation, $w = m\lambda D/d$ therefore w α D when everything else is kept constant.
(d)	The central fringe would be white and the fringes on either side of the central fringe would show the white light diffracted into its individual colours with violet being closest to the central fringe.	1 1	**2**	Remember longer wavelength (towards red end of visible spectrum) are diffracted most.

10. (a)	*Time / s* — 5·0, 10·0, 15·0, 20·0, 25·0 *p.d. / V* — 2·0, 4·0, 6·0, 8·0, 10·0 *Current / mA* — 4·0, 4·2, 3·9, 4·0, 4·0 *Charge / mC* — 20·0, 42·0, 58·5, 80·0, 100·0 Deduct 1 mark for any wrong value **once**.	**2**	**2**	
(b)	The switch is closed and the capacitor is allowed to charge whilst the variable resistor is adjusted continuously in order to maintain a constant charging current.	**1**	**3**	
	The time taken (using the stopwatch) for the p.d. across the capacitor to reach 2·0 V then 4·0 V and so on from the **uncharged** capacitor is measured.	**1**		
	The charge stored at each p.d. is found using Q = It where I is the constant charging current.	**1**		
(c)	The graph should be a best-fitting straight line passing close to the origin and the 10·0 V, 100 mC point. 1 mark for correctly plotted points and 1 mark for best fitting straight line corresponding to above. Deduct 1 mark for unlabelled axes.		**2**	The origin should be clearly labelled '0' on both axes.
(d)	$E = \frac{1}{2}QV$ $= \frac{1}{2} \times 100 \times 10^{-3} \times 10$ $= 0{\cdot}50$ J (500 mJ)	**1** **1** **1**	**3**	Area under the graph could also be used.
(e)	First find the capacitance, $C = Q/V$ $= 100 \times 10^{-3} \times 10$ $= 10 \times 10^{-3}$ F then $E = \frac{1}{2}CV^2$ $= \frac{1}{2} \times 10 \times 10^{-3} \times 5{\cdot}4^2$ $= 146$ mJ *(1) for **both** formulae Answer must be left in mJ – this is a 'show that' question.	 **1** **1** ***1** **1**	**4**	Do **not** use the graph to estimate the value of Q when $V_c = 5{\cdot}4$ V – too inaccurate.

11. (a) (i)	Electrons Arsenic has 5 electrons in its outer shell therefore there is a 'spare' electron when used as a doping agent with the silicon atoms.	1 1	2	Arsenic has a valency of 5 and silicon has a valency of 4 hence arsenic has an extra electron left over after the arsenic bonds with 4 surrounding silicon atoms. The periodic table provided in the examination would be required here for the valency of arsenic.
(ii)	It is electrically neutral. Although arsenic provides an additional 'spare' electron the arsenic nucleus also has an additional proton therefore the doped silicon is still electrically neutral.	1 1	2	
(b)	In forward bias mode if the p.d. across the diode reaches the required 'switch-on' voltage electrons will gain sufficient energy to overcome any energy difference between the valence band and the conduction band. This energy gain allows electrons to move from the valence band into the conduction band. The p-n junction diode now acts like a conductor allowing an electron current to pass from the n-type to the p-type semiconducting material.	1 1 1	3	A typical silicon diode has a 'switch on' voltage of approximately 0·7 V.

Practice Exam B

Section 1

Question	Response	Mark	Top Tips
1.	C	**1**	
2.	C	**1**	The graph shows constant velocity.
3.	B	**1**	Impulse is a vector quantity.
4.	B	**1**	
5.	B	**1**	Moving objects contract in the direction of the moving object.
6.	A	**1**	$f = f_s \left(\frac{v}{v - v_s} \right)$
7.	C	**1**	$F = \frac{G m_1 m_2}{r^2}$
8.	D	**1**	The wavelength is lengthened (red-shifted) so the galaxy is moving away – but *not* at the speed of light! $z = \frac{v}{c}$
9.	D	**1**	Charged particles travelling at right angles to field lines experience the maximum force. If a charged particle travelled exactly parallel to (or along) a field line it would experience no force.

10.	E	**1**	$^{232}_{90}\text{Th} \rightarrow {}^{228}_{90}\text{Th} + ?$ Mass reduction must be due to the emission of one α particle. $^{232}_{90}\text{Th} \rightarrow {}^{228}_{90}\text{Th} + {}^{4}_{2}\alpha$ Now to balance the proton number you need the emission of two β particles $^{232}_{90}\text{Th} \rightarrow {}^{228}_{90}\text{Th} + {}^{4}_{2}\alpha + {}^{0}_{-1}\beta + {}^{0}_{-1}\beta$
11.	C	**1**	↑ absorption ↓ emission $E = hf$. A larger energy transition means the photon has higher frequency/shorter wavelength.
12.	A	**1**	$m\lambda = d\sin\theta,\ m = 1$ $\sin\theta = \dfrac{\lambda}{d}$ To obtain smallest θ, λ should be smallest and d, grating spacing should be largest, i.e. grating lines per millimeter should be smallest value.
13.	E	**1**	$\text{Refractive Index} = \dfrac{\sin\theta_{air}}{\sin\theta_{glass}}$ $\theta_{air} = 55°$ $\theta_{glass} = 33°$
14.	E	**1**	$Id^2 = \text{a constant}$
15.	D	**1**	

16.	C	**1**	Remember $R_T = R_1 + R_2$ for series $\frac{1}{R_T} = \frac{1}{R_1} + \frac{1}{R_2}$ for parallel In a parallel arrangement the total resistance is always smaller than the smallest resistor.
17.	C	**1**	$I = \frac{V}{\text{external R}}$
18.	B	**1**	Use $T = \frac{1}{f}$ to find period and then relate this to the wavelength to find the time base.
19.	D	**1**	Increasing irradiance means *more photons* are produced. This increases the current (of photoelectrons). The colour (frequency) of this source does not change ($E = hf$) so individual photoelectrons still have the same energy.
20.	E	**1**	First calculate the absolute uncertainties then use these to find the percentage uncertainties and compare.

Practice Exam B

Section 2

Question	Expected response	Mark	Max Mark	Top Tips
1. (a)	$a=\frac{F}{m}$	1	4	Remember, F = ma.
	$=\frac{75}{100}\left(\frac{4\times348\times10^3}{522\times10^3}\right)$	2		F is the unbalanced force. In the real world, friction forces have to be considered.
	$=2\cdot0\text{ms}^{-2}$	1		
(b)	Air resistance (which increases with speed)	1	1	
(c)	$v=u+at$		3	
	$\Rightarrow t=\frac{v-u}{a}$	1		
	$=\frac{80-0}{1\cdot6}$	1		
	$=50\text{s}$	1		
(d)	Minimum length = take off run to v_1 + braking distance	1	7	
	Take off run $v^2=u^2+2as$			
	$\Rightarrow s=\frac{v^2-u^2}{2a}$	1		
	$=\frac{75^2-0}{2\times1\cdot6}$			
	$=\frac{5625}{3\cdot2}$			
	$=1758\text{m}$	1		
	Braking distance $s=\frac{v^2-u^2}{2a}$	1		
	$=\frac{0-75^2}{2\times(-3\cdot5)}$	1		
	$=804\text{m}$	1		
	$\Rightarrow$ Minimum length = 1758 + 804			
	= 2562 m	1		

2. (a)	$\Delta mv = mv - mu$ $= (0{\cdot}06 \times -4{\cdot}2) - (0{\cdot}06 \times 4{\cdot}9)$ $= -0{\cdot}546 \text{ kgms}^{-1}$	1 1 1	3	Remember, momentum is a **vector**!
(b)	$F = \frac{\Delta mv}{t}$ $= \frac{-0{\cdot}546}{0{\cdot}002}$ $= 273\text{N}$	1 1 1	3	Magnitude only required.
(c)	E_K before bounce $= \frac{1}{2}mv^2$ $= \frac{1}{2} \times 0{\cdot}06 \times (4{\cdot}9)^2$ $= 0{\cdot}72 \text{ J}$ E_K after bounce $= \frac{1}{2} \times 0{\cdot}06 \times (4{\cdot}2)^2$ $= 0{\cdot}53 \text{ J}$ $\Rightarrow E_K \text{ lost} = 0{\cdot}19 \text{ J}$ $\% E_K \text{ lost} = \frac{0{\cdot}19}{0{\cdot}72} \times 100$ $= 26{\cdot}4\%$	1 1 1 1 1 1 1 1	8	
(d) (i)	Mean value $= \frac{263}{10} = 26{\cdot}3\%$	1	1	
(ii)	Random uncertainty $= \frac{\text{max} - \text{min}}{\text{number of readings}}$ $= \frac{29{\cdot}2 - 24{\cdot}0}{10}$ $= \pm 0{\cdot}5$	1 1 1	3	
(e)	Demonstrates no understanding Limited understanding Reasonable understanding Good understanding **1 mark:** The student has demonstrated a limited understanding of the physics involved. The student has made some statement(s) which is/are relevant to the situation, showing that at least a little of the Physics within the problem is understood.	0 1 2 3	3	Open-ended question. Try and write down everything you know about the physics described in the question and, if possible, include a law of physics or equation.

	3 marks: The maximum available mark would be awarded to a student who has demonstrated a good understanding of the physics involved. The student shows a good comprehension of the physics of the situation and has provided a logically correct answer to the question posed. This type of response might include a statement of the principles involved, a relationship or an equation, and the application of these to respond to the problem. This does not mean the answer has to be what might be termed an 'excellent' answer or a 'complete' one.			
3. (a) (i)	$\frac{1}{2}CV^2 = mgh \Rightarrow h = \frac{CV^2}{2mg}$ $= \frac{1\times10^{-6}\times(5\times10^3)^2}{2\times0{\cdot}02\times9{\cdot}8}$ $= 64\text{m}$	**1** **1** **1**	**3**	When an object is raised up, it gains gravitational potential energy $E_p = mgh$
(ii)	$Pt = mgh \Rightarrow h = \frac{Pt}{mg}$ $= \frac{100\times6\times60\times60}{20\times9{\cdot}8}$ $= 11{,}020\text{m}$	**1** **1** **1**	**3**	
(b)	Demonstrates no understanding Limited understanding Reasonable understanding Good understanding **1 mark:** The student has demonstrated a limited understanding of the physics involved. The student has made some statement(s) which is/are relevant to the situation, showing that at least a little of the Physics within the problem is understood.	**0** **1** **2** **3**	**3**	Open-ended question. Try and write down everything you know about the physics described in the question and, if possible, include a law of physics or equation.

	For example, 1 mark might be awarded for the following statement: **2 marks:** The student has demonstrated a reasonable understanding of the physics involved. The student makes some statement(s) which is/are relevant to the situation, showing that the problem is understood.			Note that the answers provided here are only suggested responses; there may be more and varied answers to the question.
	3 marks: The maximum available mark would be awarded to a student who has demonstrated a good understanding of the physics involved. The student shows a good comprehension of the physics of the situation and has provided a logically correct answer to the question posed. This type of response might include a statement of the principles involved, a relationship or an equation, and the application of these to respond to the problem. This does not mean the answer has to be what might be termed an 'excellent' answer or a 'complete' one.			
4. (a) (i)	$z = \frac{\lambda observer - \lambda rest}{\lambda rest} = \frac{v}{c}$ $\Rightarrow v = \frac{(\lambda observer - \lambda rest)c}{\lambda rest}$ $= \frac{(396 \cdot 2 - 393 \cdot 4) \times 3 \times 10^8}{393 \cdot 4}$ $= 2{\cdot}1 \times 10^6$ ms^{-1} (2100 kms^{-1})	1 1 1 1	4	
(ii)	The wavelength is longer. It is shifted towards the red (longer wavelength) end of the visible spectrum. The galaxy is receding from Earth.	1 1	2	Make sure you understand the terms 'redshift' and 'blueshift'.

(b)	Graph similar to below.	**3**	**3**	1 mark for labels. 1 mark for correctly plotted data. 1 mark for best fitting straight line.
(c)	V, d The value of H_0 can be obtained from the **gradient**. $y = mx$ $v = H_0 d$ gradient	**1**	**1**	
(d)	$m = \frac{y_2 - y_1}{x_2 - x_1} = \frac{5990 - 2100}{(26 \cdot 2 - 8 \cdot 60) \times 10^{20}}$ gradient $m = H_o = \frac{3890}{17 \cdot 6 \times 10^{20}}$ $= 2{\cdot}21 \times 10^{-18}\ s^{-1}$	**1** **1** **1**	**3**	Note: if origin is used the gradient will be $2.29 \times 10^{-18}\ s^{-1}$
5. (a)	$V = 4 \times 2$ $= 8V.$	**1**	**1**	Make sure you understand the Y gain and timebase settings on an oscilloscope.
(b)	Timebase set at 5 ms cm^{-1} Time for one complete charge and discharge. $t = 5 \times 4$ $= 20$ ms $f = \frac{1}{T} = \frac{1}{20 \times 10^{-3}}$ $= 50$Hz	**1** **1** **1**	**3**	
(c)	$C = \frac{Q}{V} = \frac{It}{V} = \frac{40 \times 10^{-3} \times 20 \times 10^{-3}}{8}$ $= 1 \times 10^{-4}$F (or 100μF)	**1** **1** **1**	**3**	
6. (a) (i)	$\lambda_{air} = \frac{v}{f} = \frac{3 \times 10^8}{4 \cdot 3 \times 10^{14}}$ $= 7 \cdot 0 \times 10^{-7}$ m	**1** **1** **1**	**3**	
(ii)	$\lambda_{glass} = \frac{\lambda_{air}}{n} = \frac{7 \cdot 0 \times 10^{-7}}{1 \cdot 94}$ $= 3 \cdot 6 \times 10^{-7}$ m	**1** **1** **1**	**3**	

(b) (i)	For red light, $\sin c = \frac{1}{n} = \frac{1}{1 \cdot 94}$ $\Rightarrow c = 31°$ $\theta_{glass} = 30°$ therefore the critical angle is not exceeded. The red beam A is refracted as shown. For violet light $\sin c$ $= \frac{1}{2 \cdot 06} \Rightarrow c = 29°$ The critical angle is exceeded so total internal reflection occurs as shown.	1 1 1 1	4	Remember, total internal reflection occurs when the angle of incidence **exceeds** the critical angle.
(ii)	$\frac{\sin\theta_{air}}{\sin\theta_{glass}} = 1 \cdot 94$ $\Rightarrow \sin\theta_{air} = 1 \cdot 94 \sin\theta_{glass}$ $= 1 \cdot 94 \sin 30°$ $\Rightarrow \theta_{air} = 75 \cdot 9°$ For reflection, $i = r \Rightarrow z = 30°$	1 1 1	3	
7. (a)	$n\lambda = d \sin\theta \quad (n = 1)$. $\sin\theta = \frac{n\lambda}{d} = n\lambda\left(\frac{1}{d}\right)$ $\frac{1}{d} = 6 \times 10^5$ $\sin\theta = 410 \times 10^{-9} \times 6 \times 10^5$ $= 0 \cdot 246$ $\Rightarrow \theta = x = 14 \cdot 2°$	1 1 1 1	4	Grating space (d) in metres $= \frac{1}{\text{Number of lines per metre}}$
(b)	The order of colours would be reversed. OR Unlike a grating, a prism produces a single spectrum.	1	1	
8. (a)	Proton, u u d π^-, $\bar{u}$ d (where $\bar{u}$ is an antiquark)	 1 1	2	Remember, antiparticles have the opposite charge.

(b)	(i)	$s = vt = 0{\cdot}9995 \times 3 \times 10^8 \times 2{\cdot}2 \times 10^{-6}$ $= 659{\cdot}7$ m	1 1 1	3	
	(ii)	$t' \frac{t}{\sqrt{1-\left(\frac{v}{c}\right)^2}} = \frac{2{\cdot}2\times10^{-6}}{\sqrt{1-(0{\cdot}9995)^2}}$ $= \underline{6{\cdot}96} \times \underline{10^{-5}}\ \underline{s}$ distance travelled by muons = vt' $= 0{\cdot}9995 \times 3 \times 10^8 \times 6{\cdot}96 \times 10^{-5}$ = 20,870 m (20·9 km) If the muons were created at an altitude of 15km then they will reach the Earth before they decay.	1 1 1 1 1 1	6	
9. (a)		Negative values describe a 'bound state' where electrons are held in orbits. In order to achieve ionisation (where an electron is freed from the atom), the electron must be given (positive) energy. E.g. an electron in level 5 would need $8{\cdot}71 \times 10^{-20}$ J of energy to free it.	1 1	2	
(b)	(i)	The longest wavelength corresponds to the lowest frequency and therefore the lowest energy. (from E_3 to E_2) $\lambda = \frac{ch}{E}$ $f = \frac{E}{h}$	1	1	
	(ii)	$\lambda = \frac{c}{f}$, $f = \frac{E}{h}$ } 1 for both $\Rightarrow \lambda = \frac{ch}{E}$ $= \frac{3\times10^8\times6{\cdot}63\times10^{-34}}{(54{\cdot}2-24{\cdot}2)\times10^{-20}}$ $= 6{\cdot}63\times10^{-7}$ m (663nm)	1 1 1 1	4	

10. (a)	$E_P = E_K$ $mgh = \frac{1}{2} mv^2$ $\Rightarrow gh = \frac{1}{2} v^2$ $\Rightarrow v^2 = 2gh$	1 1	2	
(b)	The graph should be 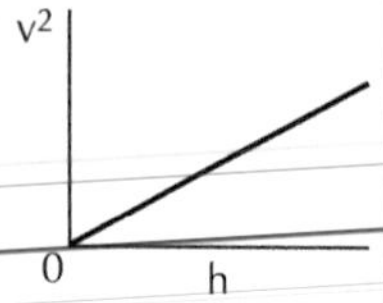 The gradient of the graph = 2g. Divide your gradient by 2 to get the value of g.	1 1 1 1	4	1 for labelling. 1 for straight line through origin.
(c) (i)	A systematic uncertainty	1	1	
(ii)	All values of h would be too large (by the same amount). The graph would no longer pass through the origin but the gradient and therefore the value of g would not be affected.	1 1	2	

Practice Exam C

Section 1

Question	Response	Mark	Top Tips
1.	A	**1**	Remember that 'g' and 'u' have opposite signs and consider the fact that the sandbag is still **rising** when it is released.
2.	C	**1**	Displacement is a vector. Remember to subtract areas.
3.	B	**1**	Ave force = change in momentum/time Uncertainty in ave force – take the largest percentage uncertainty from all the quantities used.
4.	A	**1**	You need the gravitational force equation here.
5.	C	**1**	Remember t' = time in the stationary frame of reference and t is the time according to the moving frame of reference.
6.	D	**1**	The lowest frequency is detected when the siren is moving away and the highest when the siren is moving towards the observer.
7.	E	**1**	Increasing the intensity increases the number of photons arriving not their energy – that is controlled by the frequency of the photons.
8.	E	**1**	The idea of "cut off frequency" might give you a clue here.
9.	B	**1**	Standard use of Snell's law twice. Don't forget, $i = 35°$
10.	A	**1**	Inverse square law for irradiance.
11.	A	**1**	Knowledge relating to the different quark constituents of mesons & baryons required. Fermions have 6 not 3 types of quarks and leptons. Understand what is meant by a force-mediating particle.
12.	D	**1**	Use $\sin c = 1/n$ to calculate the critical angle. Remember total internal reflection only occurs when angle of incidence is greater than the critical angle.
13.	E	**1**	Electric field direction is defined in terms of the direction in which a POSITIVE test charge would move if placed near another charge.
14.	B	**1**	This is a parallel circuit: $1/R = 1/(3 + 1) + 1/(5 + 7) = 1/4 + 1/12$
15.	D	**1**	Negative charges use the **right** hand rule. First finger – mag field direction, second finger – velocity direction and thumb – direction of motion.
16.	D	**1**	$E = IR + Ir$ is useful here

17.	B	**1**	Resistances add irrespective of battery orientation, Emf's in opposition subtract.
18.	A	**1**	Remember $V_{rms} = V_{peak}/\sqrt{2}$
19.	E	**1**	The more resistances in parallel the **smaller** the total resistance and hence the greater the current.
20.	D	**1**	E = hf and E = qV are both needed here

Practice Exam C

Section 2

Question	Expected response	Mark	Max Mark	Top Tips
1. (a) (i)	$u_H = u \cos \theta$ $= 7{\cdot}2 \cos 50°$ $= 4{\cdot}6\ ms^{-1}$	 1 	1	Standard opening to a projectile question.
(ii)	$u_v = u \sin \theta$ $= 7{\cdot}2 \sin 50°$ $= 5{\cdot}5\ ms^{-1}$	 1 	1	Standard opening to a projectile question.
(b)	Time from A→B $v = u + at$ (vert) $0 = 5{\cdot}5 + (-9{\cdot}8)t$ $t = 0{\cdot}56\ s$ Time from B → C $s = ut + ½ at^2$ (vert) $1{\cdot}02 = 0 + 4{\cdot}9t^2$ $t = 0{\cdot}46\ s$ Therefore total time $= 0{\cdot}56 + 0{\cdot}46$ $= 1{\cdot}02\ s$	 1 1 1 1 1 1 1	7	This problem is most easily solved by splitting the motion up into two simpler projectiles and calculating the time involved for each before adding them.
(c)	(horiz) $s_H = u_H \times t$ $= 4{\cdot}6 \times 1{\cdot}02$ $= 4{\cdot}7\ m$	 1 1 1	3	
2. (a)	$F = ma.$ $m = (56{,}000 + 250 \times 80) = 76{,}000\ kg$ $a = \frac{v-u}{t} = \frac{12-0}{10} = 1{\cdot}2\ ms^{-2}$ $F = 76{,}000 \times 1{\cdot}2$ $= 91{\cdot}2\ kN$	1 1 1 1 1	5	

(b)	$P = \frac{E}{t} = \frac{\frac{1}{2}mv^2}{t}$ $= \frac{\frac{1}{2} \times 76{,}000 \times 12^2}{10}$ $= 547$ kW	1 1 1	3	The moving tram has kinetic energy $Ek = \frac{1}{2}mv^2$
(c)	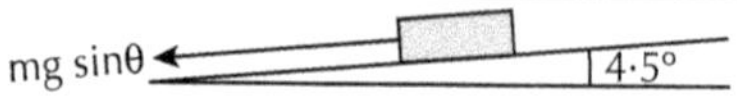 mg sinθ ← [4·5°] The extra force is mg $\sin\theta$ $F = 76{,}000 \times 9{\cdot}8 \times \sin 4{\cdot}5°$ $= 58{\cdot}4$ kN	1 1 1	3	Remember the force down a slope is: mg $\sin\theta$
(d)	$P = VI \Rightarrow I = \frac{P}{V} = \frac{547{,}200}{750}$ $= 729{\cdot}6$ A $= 730$ A	1 1 1	3	
3.	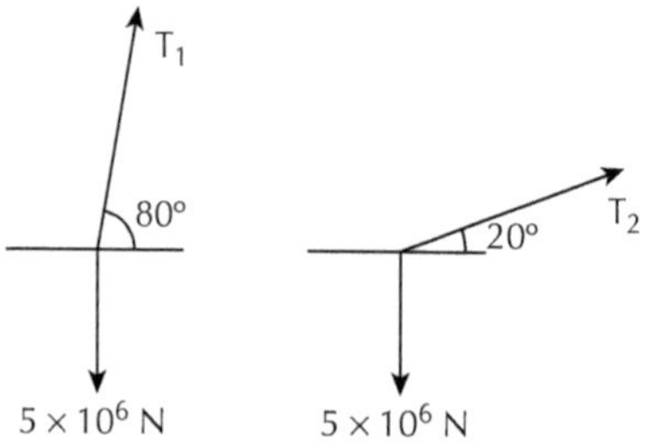 T_1, 80°, 5×10^6 N; T_2, 20°, 5×10^6 N Remembering that the 10^7N load is supported equally by two cables, the vertical load per cable is $\frac{1}{2} \times 1 \times 10^7 = 5 \times 10^6$N $T_1 \sin 80° = 5 \times 10^6 \Rightarrow T_1 = \frac{5 \times 10^6}{\sin 80°}$ $= 5{\cdot}1 \times 10^6$N $T_2 \sin 20° = 5 \times 106 \Rightarrow T_2 = \frac{5 \times 10^6}{\sin 20°}$ $= 1{\cdot}46 \times 10^7$N The tensions should range from $5{\cdot}1 \times 10^6$ N to a maximum of $1{\cdot}46 \times 10^7$ N.	1 1 1 1 1	5	This is an example of resolution of a force into components 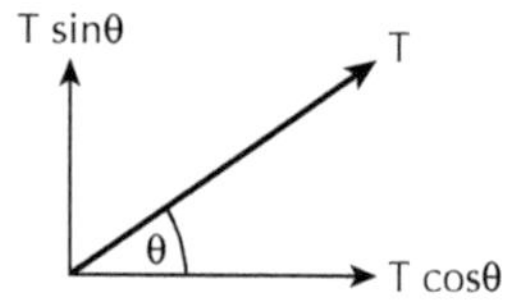 T sinθ, T, θ, T cosθ In this case it is the vertical component which supports the (vertical) load.

4. (a)	Demonstrates no understanding Limited understanding Reasonable understanding Good understanding **1 mark:** The student has demonstrated a limited understanding of the physics involved. The student has made some statement(s) which is/are relevant to the situation, showing that at least a little of the Physics within the problem is understood. **2 marks:** The student has demonstrated a reasonable understanding of the physics involved. The student makes some statement(s) which is/are relevant to the situation, showing that the problem is understood. **3 marks:** The maximum available mark would be awarded to a student who has demonstrated a good understanding of the physics involved. The student shows a good comprehension of the physics of the situation and has provided a logically correct answer to the question posed. This type of response might include a statement of the principles involved, a relationship or an equation, and the application of these to respond to the problem. This does not mean the answer has to be what might be termed an 'excellent' answer or a 'complete' one.	**0** **1** **2** **3**	**3**	Open-ended question. Try and write down everything you know about the physics described in the question and, if possible, include a law of physics or equation.

(b)	The astronaut gains momentum as a result of the push on the Space Station in the opposite direction. Newton's third law states that at the same time the Space Station gains the same amount of momentum as the astronaut in the opposite direction. Hence the velocity of both changes.	1 1	**2**	
(c)	The change in momentum of the gas is balanced by an equal and opposite change of momentum of the Space Station resulting in a velocity which can change its direction.	1 1	**2**	
5. (a)	The astronaut will be younger than his Earthbound twin because of time dilation. Using the equation and taking a time on Earth as 10 years $t' = \frac{t}{\sqrt{1 - \frac{v^2}{c^2}}}$ $10 = \frac{t}{\sqrt{1 - \left(\frac{2 \cdot 6 \times 10^8}{3 \times 10^8}\right)^{22}}}$ We obtain $t = 5$ years.	1 1 1 1	**4**	'Moving clocks' run slow. The astronaut only ages 5 years during the 10 years experienced by his physicist brother.

(b)	The speed of light is the same for all observers. Both experiments will yield the same result.	 1 1	2	
(c)	Demonstrates no understanding Limited understanding Reasonable understanding Good understanding **1 mark:** The student has demonstrated a limited understanding of the physics involved. The student has made some statement(s) which is/are relevant to the situation, showing that at least a little of the Physics within the problem is understood. **2 marks:** The student has demonstrated a reasonable understanding of the physics involved. The student makes some statement(s) which is/are relevant to the situation, showing that the problem is understood. **3 marks:** The maximum available mark would be awarded to a student who has demonstrated a good understanding of the physics involved. The student shows a good comprehension of the physics of the situation and has provided a logically correct answer to the question posed. This type of response might include a statement of the principles involved, a relationship or an equation, and the application of these to respond to the problem. This does not mean the answer has to be what might be termed an 'excellent' answer or a 'complete' one.	0 1 2 3	3	Open-ended question. Try and write down everything you know about the physics described in the question and, if possible, include a law of physics or equation.

6.	(a)	An antiparticle has the same mass and opposite charge to its corresponding particle.	1	1	
	(b)	A meson is composed of a quark and an antiquark.	1	1	Know the composition of mesons and baryons.
	(c)	An up (u) quark and an antidown (d) quark.	1 1	2	Up quark charge = $+\frac{2}{3}$e Antidown quark charge = $+\frac{1}{3}$e Total charge = +1e
	(d)	(Virtual) photons.	1	1	You need to be aware of the force mediating particle associated with different forces.
	(e)	Infinite range	1	1	You need to be aware of the properties associated with each type of force (strong, weak, electromagnetic and gravitational).
	(f)	β-particle decay	1	1	A neutron can undergo decay to produce a proton, an electron (β-particle) and an antineutrino. The existence of antineutrinos suggests the existence of neutrinos.
7.	(a)	p.d. = 200 kv	1	1	
	(b) (i)	The strength of the electric field is the same at all points between the plates.	1	1	
	(ii)	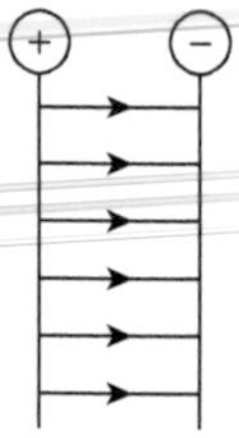 + −	2	2	1 for equally spaced lines. 1 for direction of arrows.

	(c) $\frac{1}{2}mv^2 = \frac{1}{2}mu^2 + qV$ $v^2 = \frac{mu^2 + 2qV}{m}$ $v = \sqrt{\frac{mu^2 + 2qV}{m}}$ $= \sqrt{\frac{6 \cdot 65 \times 10^{-27} \times (2 \cdot 53 \times 10^6)^2 + 2 \times 3 \cdot 2 \times 10^{-19} \times 2 \times 10^5}{6 \cdot 65 \times 10^{-27}}}$ $\Rightarrow v = 5 \cdot 06 \times 10^6 \text{ms}^{-1}$	1 1 1 1	4	
	(d) $F = ma$ $v^2 = u^2 + 2as$ $\Rightarrow a = \frac{v^2 - u^2}{2s}$ $\Rightarrow F = m\left(\frac{v^2 - u^2}{2s}\right)$ $= \frac{6 \cdot 65 \times 10^{-27}\left[\left(5 \cdot 06 \times 10^6\right)^2 - \left(2 \cdot 53 \times 10^6\right)^2\right]}{2 \times 0 \cdot 2}$ $= 3.2 \times 10^{-13} \text{N}$	1 1 1	3	
8. (a)	Nuclear fission is the splitting up of a heavy nucleus to produce two lighter nuclei and by-products such as neutrons, electrons and heat energy.	1	1	
(b)	Mass of LHS $1{\cdot}673 \times 10^{-27} + 390{\cdot}219 \times 10^{-27}$ $= 3{\cdot}91892 \times 10^{-25}$ kg Mass of RHS $227{\cdot}291 \times 10^{-27} + 157{\cdot}562 \times 10^{-27} +$ $(4 \times 1{\cdot}673 \times 10^{-27})$ $= 3{\cdot}91545 \times 10^{-25}$ kg Mass difference $= 3{\cdot}47 \times 10^{-28}$ kg $E = mc^2$ $= 3{\cdot}47 \times 10^{-28} \times (3{\cdot}00 \times 10^8)^2$ $= 3{\cdot}12 \times 10^{-11}$ J	1 1 1 1	4	**Never** round any numbers until the very end!

(c)	Number of fissions = Total energy per second (power)/energy of each fission reaction. $= 600 \times 10^{6} / 3{\cdot}12 \times 10^{-11}$ $= 1{\cdot}92 \times 10^{19}$ fission reactions	 1 1 1	**3**	
(d)	**Any 2 from:** Temperature: Fusion requires high temperatures (high activation energy - estimated to be 40 million Kelvin) to get the nuclear fusion reaction started. Heat is used to provide the energy, but it takes a lot of heat to start the reaction. and/or Time: Charged nuclei must be held together close enough and long enough for the fusion reaction to start. Scientists estimate that the plasma needs to be held together at 40,000,000 K for about one second. and/or Containment: Containment of the plasma within the reaction chamber is a major problem since there are no materials capable of withstanding the extremely high temperatures associated with fusion. Since the plasma is charged, magnetic fields are used to try and contain it but it is very difficult to prevent the confinement field from leaking plasma and, as soon as this happens, the fusion process stops.	1 1 1 1 1 1	**4**	

9.	$z = \frac{\lambda_{observer} - \lambda_{rest}}{\lambda_{rest}} = \frac{v}{c}$ $\Rightarrow \frac{50}{650} = \frac{v}{3 \times 10^8}$ $\Rightarrow v = 2{\cdot}3 \times 10^7$ ms^{-1} This exceeds the speed limit of 2×10^7 ms^{-1}. The judge knew his Physics!	1 1 1 1	4	This is the Doppler Effect. You have probably seen more examples of the source moving and the observer stationary but the equation also works when the observer is moving towards a stationary light source provided his speed is significantly less than the speed of light.
10. (a)	Microwaves from the transmitter meet waves which have been reflected off the aluminium and are travelling back towards the transmitter. Microwaves that are reflected from the reflector interfere constructively and destructively producing an interference pattern. Where the wavecrest of an incident wave (to the receiver) meets a wavecrest from the reflected waves they interfere constructively producing a large voltage reading on the voltmeter. Where the wavetrough of an incident wave (to the receiver) meets a wavecrest from the reflected waves they interfere destructively producing a low voltage reading on the voltmeter.	1 1 1 1 1	5	The type of interference produced is called a standing wave. The terms in-phase or out-of-phase may be used here as an alternative. The terms in-phase and out-of-phase may be used here as an alternative.
(b) (i)	The receiver moves through a total of 11 points of constructive interference.	1	1	Remember the receiver starts and finishes on a point of constructive interference.

(ii)	$1\lambda = 3$ points of constructive interference 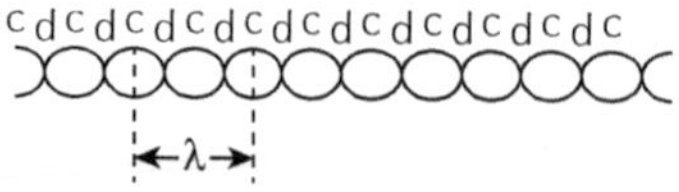 c d c d c d c d c d c d c d c d c d c d c ←λ→ 11 points of constructive interference $= 5\lambda$ $5\lambda = 140$ mm, $\lambda = 28$ mm	1 1 1	3	Number of wavelengths = (Number of points of constructive interference −1)/2.
11. (a)	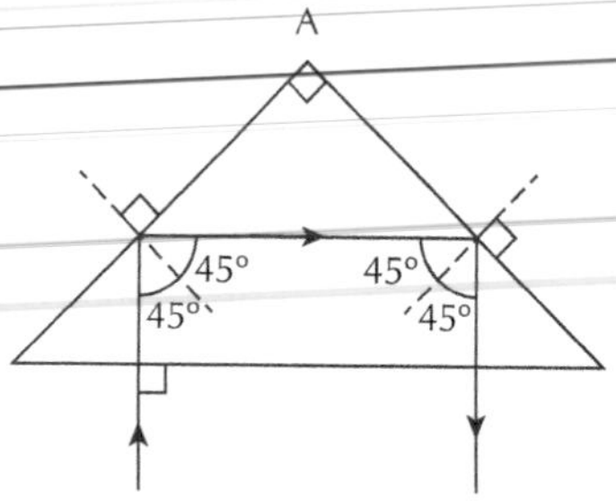 A 45° 45° 45° 45° $\sin c = \frac{1}{n} = \frac{1}{1 \cdot 45}$ $\Rightarrow c = 43 \cdot 6^\circ$ Total internal reflection will occur (twice) in A 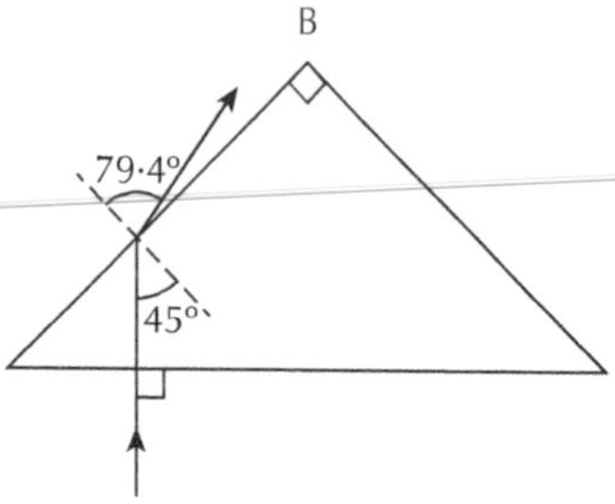 B 79·4° 45° $\left(\sin c = \frac{1}{n} = \frac{1}{1 \cdot 39} \Rightarrow c = 46^\circ\right)$ $\frac{\sin\theta_{air}}{\sin\theta_{glass}} = 1 \cdot 39$ Exiting B $\Rightarrow \frac{\sin\theta_{air}}{\sin 45} = 1 \cdot 39$ Sin $\theta = 1{\cdot}39 \sin 45$ $\Rightarrow \theta_{air} = 79{\cdot}4^\circ$	1 1 1 1 1 1 1 1 1	9	You must check the critical angles $\sin c = \frac{1}{n}$ then, if refraction occurs $\frac{\sin\theta_{air}}{\sin\theta_{glass}} = n$ $n = \frac{\lambda_{air}}{\lambda_{glass}}$

(b)	Prism A $\frac{\lambda_{air}}{\lambda_{glass}} = 1 \cdot 45 \Rightarrow \frac{650}{\lambda_{glass}} = 1 \cdot 45$ $\Rightarrow \lambda_{glass} = 448 \cdot 3\text{nm}$ Prism B $\frac{\lambda_{air}}{\lambda_{glass}} = 1 \cdot 39 \Rightarrow \frac{650}{\lambda_{glass}} = 1 \cdot 39$ $\Rightarrow \lambda_{glass} = 467 \cdot 6\text{nm}$ $\Rightarrow$ Difference = 467·6 – 448·3 = 19·3 nm	1 1 1 1	**4**	
12. (a)	Since the power supply is an a.c. generator the capacitor will charge up, discharge and then re-charge in the opposite direction again. Electrons move on and off the capacitor plates at the same frequency as the supply. The charging and discharging currents pass through the lamp continuously ensuring the lamp stays lit.	1 1 1	**3**	
(b)	At a higher frequency, more charge per second (current) passes through the lamp producing a hotter filament and hence brighter lamp.	1 1	**2**	
(c)	The lamp will get dimmer. A smaller capacitor stores less charge/energy during each charging and discharging cycle.	1 1	**2**	

DATA SHEET

COMMON PHYSICAL QUANTITIES

Quantity	Symbol	Value	Quantity	Symbol	Value
Speed of light in vacuum	c	$3{\cdot}00 \times 10^{8}$ ms^{-1}	Planck's constant	h	$6{\cdot}63 \times 10^{-34}$ Js
			Mass of electron	m_e	$9{\cdot}11 \times 10^{-31}$ kg
Magnitude of the charge on an electron	e	$1{\cdot}60 \times 10^{-19}$C			
Universal Constant of Gravitation	G	$6{\cdot}67 \times 10^{-11}$ m^{3} kg^{-1} s^{-2}	Mass of neutron	m_n	$1{\cdot}675 \times 10^{-27}$ kg
Gravitational acceleration of Earth	g	$9{\cdot}8$ ms^{-2}	Mass of proton	m_p	$1{\cdot}673 \times 10^{-27}$ kg
Hubble's constant	H_0	$2{\cdot}3 \times 10^{-18}$ s^{-1}			

REFRACTIVE INDICES

The refractive indices refer to sodium light of wavelength 589 nm and to substances at a temperature of 273 K.

Substance	Refractive index	Substance	Refractive index
Diamond	2·42	Water	1·33
Crown glass	1·50	Air	1·00

SPECTRAL LINES

Element	Wavelength /nm	Colour	Element	Wavelength /nm	Colour
Hydrogen	656	Red	Cadmium	644	Red
	486	Blue-green		509	Green
	434	Blue-violet		480	Blue
	410	Violet	Lasers		
	397	Ultraviolet	Element	Wavelength /nm	Colour
	389	Ultraviolet	Carbon dioxide	9550 } 10590 }	Infrared
Sodium	589	Yellow	Helium-neon	633	Red

PROPERTIES OF SELECTED MATERIALS

Substance	Density /kg m^{-3}	Melting Point /K	Boiling Point /K
Aluminium	$2{\cdot}70 \times 10^{3}$	933	2623
Copper	$8{\cdot}96 \times 10^{3}$	1357	2853
Ice	$9{\cdot}20 \times 10^{2}$	273	....
Sea Water	$1{\cdot}02 \times 10^{3}$	264	377
Water	$1{\cdot}00 \times 10^{3}$	273	373
Air	1·29	......	.....
Hydrogen	$9{\cdot}0 \times 10^{-2}$	14	20

The gas densities refer to a temperature of 273 K and a pressure of $1{\cdot}01 \times 10^{5}$pa.

Relationships required for Physics Higher

$d = \bar{v}t$

$s = \bar{v}t$

$v = u + at$

$s = ut + \frac{1}{2}at^2$

$v^2 = u^2 + 2as$

$s = \frac{1}{2}(u + v)t$

$W = mg$

$F = ma$

$E_W = Fd$

$E_P = mgh$

$E_k = \frac{1}{2}mv^2$

$P = \frac{E}{t}$

$p = mv$

$Ft = mv - mu$

$F = \frac{Gm_1m}{r^2}$

$t' = \frac{t}{\sqrt{1 - (v/c)^2}}$

$l' = l\sqrt{1 - (v/c)^2}$

$f_o = f_s\left(\frac{v}{v \pm v_s}\right)$

$z = \frac{\lambda_{observer} - \lambda_{rest}}{\lambda_{rest}}$

$z = \frac{v}{c}$

$v = H_0d$

$W = QV$

$E = mc^2$

$E = hf$

$E_k = hf - hf_0$

$E_2 - E_1 = hf$

$T = \frac{1}{f}$

$v = f\lambda$

$d\sin\theta = m\lambda$

$n = \frac{\sin\theta_1}{\sin\theta_2}$

$\frac{\sin\theta_1}{\sin\theta_2} = \frac{\lambda_1}{\lambda_2} = \frac{v_1}{v_2}$

$\sin\theta_c = \frac{1}{n}$

$I = \frac{k}{d^2}$

$I = \frac{P}{A}$

Path difference $= m\lambda$ or $\left(m + \frac{1}{2}\right)\lambda$ where $m = 0, 1, 2\ldots$

random uncertainty $= \frac{\text{max.value} - \text{min.value}}{\text{number of values}}$

$V_{peak} = \sqrt{2}V_{rms}$

$I_{peak} = \sqrt{2}I_{rms}$

$Q = It$

$V = IR$

$P = IV = I^2R = \frac{V^2}{R}$

$R_T = R_1 + R_2 + \ldots$

$\frac{1}{R_T} = \frac{1}{R_1} + \frac{1}{R_2} + \ldots$

$E = V + Ir$

$V_1 = \left(\frac{R_1}{R_1 + R_2}\right)V_s$

$\frac{V_1}{V_2} = \frac{R_1}{R_2}$

$C = \frac{Q}{V}$

$E = \frac{1}{2}QV = \frac{1}{2}CV^2 = \frac{1}{2}\frac{Q^2}{C}$

Additional Relationships

Circle

circumference $= 2\pi r$

area $= \pi r^2$

Sphere

area $= 4\pi r^2$

volume $= \frac{4}{3}\pi r^3$

Trigonometry

$\sin\theta = \frac{\text{opposite}}{\text{hypotenuse}}$

$\cos\theta = \frac{\text{adjacent}}{\text{hypotenuse}}$

$\tan\theta = \frac{\text{opposite}}{\text{adjacent}}$

$\sin^2\theta + \cos^2\theta = 1$

PERIODIC TABLE

Electron Arrangements of Elements

Key

Atomic number Symbol Electron arrangement Name

Transition Elements (groups 3–12)

Group 1 (1)	Group 2 (2)	(3)	(4)	(5)	(6)	(7)	(8)	(9)	(10)	(11)	(12)	Group 3 (13)	Group 4 (14)	Group 5 (15)	Group 6 (16)	Group 7 (17)	Group 0 (18)
1 H 1 Hydrogen																	2 He 2 Helium
3 Li 2,1 Lithium	4 Be 2,2 Beryllium											5 B 2,3 Boron	6 C 2,4 Carbon	7 N 2,5 Nitrogen	8 O 2,6 Oxygen	9 F 2,7 Fluorine	10 Ne 2,8 Neon
11 Na 2,8,1 Sodium	12 Mg 2,8,2 Magnesium											13 Al 2,8,3 Aluminium	14 Si 2,8,4 Silicon	15 P 2,8,5 Phosphorus	16 S 2,8,6 Sulfur	17 Cl 2,8,7 Chlorine	18 Ar 2,8,8 Argon
19 K 2,8,8,1 Potassium	20 Ca 2,8,8,2 Calcium	21 Sc 2,8,9,2 Scandium	22 Ti 2,8,10,2 Titanium	23 V 2,8,11,2 Vanadium	24 Cr 2,8,13,1 Chromium	25 Mn 2,8,13,2 Manganese	26 Fe 2,8,14,2 Iron	27 Co 2,8,15,2 Cobalt	28 Ni 2,8,16,2 Nickel	29 Cu 2,8,18,1 Copper	30 Zn 2,8,18,2 Zinc	31 Ga 2,8,18,3 Gallium	32 Ge 2,8,18,4 Germanium	33 As 2,8,18,5 Arsenic	34 Se 2,8,18,6 Selenium	35 Br 2,8,18,7 Bromine	36 Kr 2,8,18,8 Krypton
37 Rb 2,8,18,8,1 Rubidium	38 Sr 2,8,18,8,2 Strontium	39 Y 2,8,18,9,2 Yttrium	40 Zr 2,8,18, 10,2 Zirconium	41 Nb 2,8,18, 12,1 Niobium	42 Mo 2,8,18,13, 1 Molybdenum	43 Tc 2,8,18,13, 2 Technetium	44 Ru 2,8,18,15, 1 Ruthenium	45 Rh 2,8,18,16, 1 Rhodium	46 Pd 2,8,18, 18,0 Palladium	47 Ag 2,8,18, 18,1 Silver	48 Cd 2,8,18, 18,2 Cadmium	49 In 2,8,18, 18,3 Indium	50 Sn 2,8,18, 18,4 Tin	51 Sb 2,8,18, 18,5 Antimony	52 Te 2,8,18, 18,6 Tellurium	53 I 2,8,18, 18,7 Iodine	54 Xe 2,8,18, 18,8 Xenon
55 Cs 2,8,18,18, 8,1 Caesium	56 Ba 2,8,18,18, 8,2 Barium	57 La 2,8,18,18, 9,2 Lanthanum	72 Hf 2,8,18,32, 10,2 Hafnium	73 Ta 2,8,18, 32,11,2 Tantalum	74 W 2,8,18,32, 12,2 Tungsten	75 Re 2,8,18,32, 13,2 Rhenium	76 Os 2,8,18,32, 14,2 Osmium	77 Ir 2,8,18,32, 15,2 Iridium	78 Pt 2,8,18,32, 17,1 Platinum	79 Au 2,8,18, 32,18,1 Gold	80 Hg 2,8,18, 32,18,2 Mercury	81 Tl 2,8,18, 32,18,3 Thallium	82 Pb 2,8,18, 32,18,4 Lead	83 Bi 2,8,18, 32,18,5 Bismuth	84 Po 2,8,18, 32,18,6 Polonium	85 At 2,8,18, 32,18,7 Astatine	86 Rn 2,8,18, 32,18,8 Radon
87 Fr 2,8,18,32, 18,8,1 Francium	88 Ra 2,8,18,32, 18,8,2 Radium	89 Ac 2,8,18,32, 18,9,2 Actinium	104 Rf 2,8,18,32, 32,10,2 Rutherfordium	105 Db 2,8,18,32, 32,11,2 Dubnium	106 Sg 2,8,18,32, 32,12,2 Seaborgium	107 Bh 2,8,18,32, 32,13,2 Bohrium	108 Hs 2,8,18,32, 32,14,2 Hassium	109 Mt 2,8,18,32, 32,15,2 Meitnerium	110 Ds 2,8,18,32, 32,17,1 Darmstadtium	111 Rg 2,8,18,32, 32,18,1 Roentgenium	112 Cn 2,8,18,32, 32,18,2 Copernicium						

Lanthanides	57 La 2,8,18, 18,9,2 Lanthanum	58 Ce 2,8,18, 20,8,2 Cerium	59 Pr 2,8,18,21, 8,2 Praseodymium	60 Nd 2,8,18,22, 8,2 Neodymium	61 Pm 2,8,18,23, 8,2 Promethium	62 Sm 2,8,18,24, 8,2 Samarium	63 Eu 2,8,18,25, 8,2 Europium	64 Gd 2,8,18,25, 9,2 Gadolinium	65 Tb 2,8,18,27, 8,2 Terbium	66 Dy 2,8,18,28, 8,2 Dysprosium	67 Ho 2,8,18,29, 8,2 Holmium	68 Er 2,8,18,30, 8,2 Erbium	69 Tm 2,8,18,31, 8,2 Thulium	70 Yb 2,8,18,32, 8,2 Ytterbium	71 Lu 2,8,18,32, 9,2 Lutetium
Actinides	89 Ac 2,8,18,32, 18,9,2 Actinium	90 Th 2,8,18,32, 18,10,2 Thorium	91 Pa 2,8,18,32, 20,9,2 Protactinium	92 U 2,8,18,32, 21,9,2 Uranium	93 Np 2,8,18,32, 22,9,2 Neptunium	94 Pu 2,8,18,32, 24,8,2 Plutonium	95 Am 2,8,18,32, 25,8,2 Americium	96 Cm 2,8,18,32, 25,9,2 Curium	97 Bk 2,8,18,32, 27,8,2 Berkelium	98 Cf 2,8,18,32, 28,8,2 Californium	99 Es 2,8,18,32, 29,8,2 Einsteinium	100 Fm 2,8,18,32, 30,8,2 Fermium	101 Md 2,8,18,32, 31,8,2 Mendelevium	102 No 2,8,18,32, 32,8,2 Nobelium	103 Lr 2,8,18,32, 32,9,2 Lawrencium

ADDITIONAL GRAPH PAPER

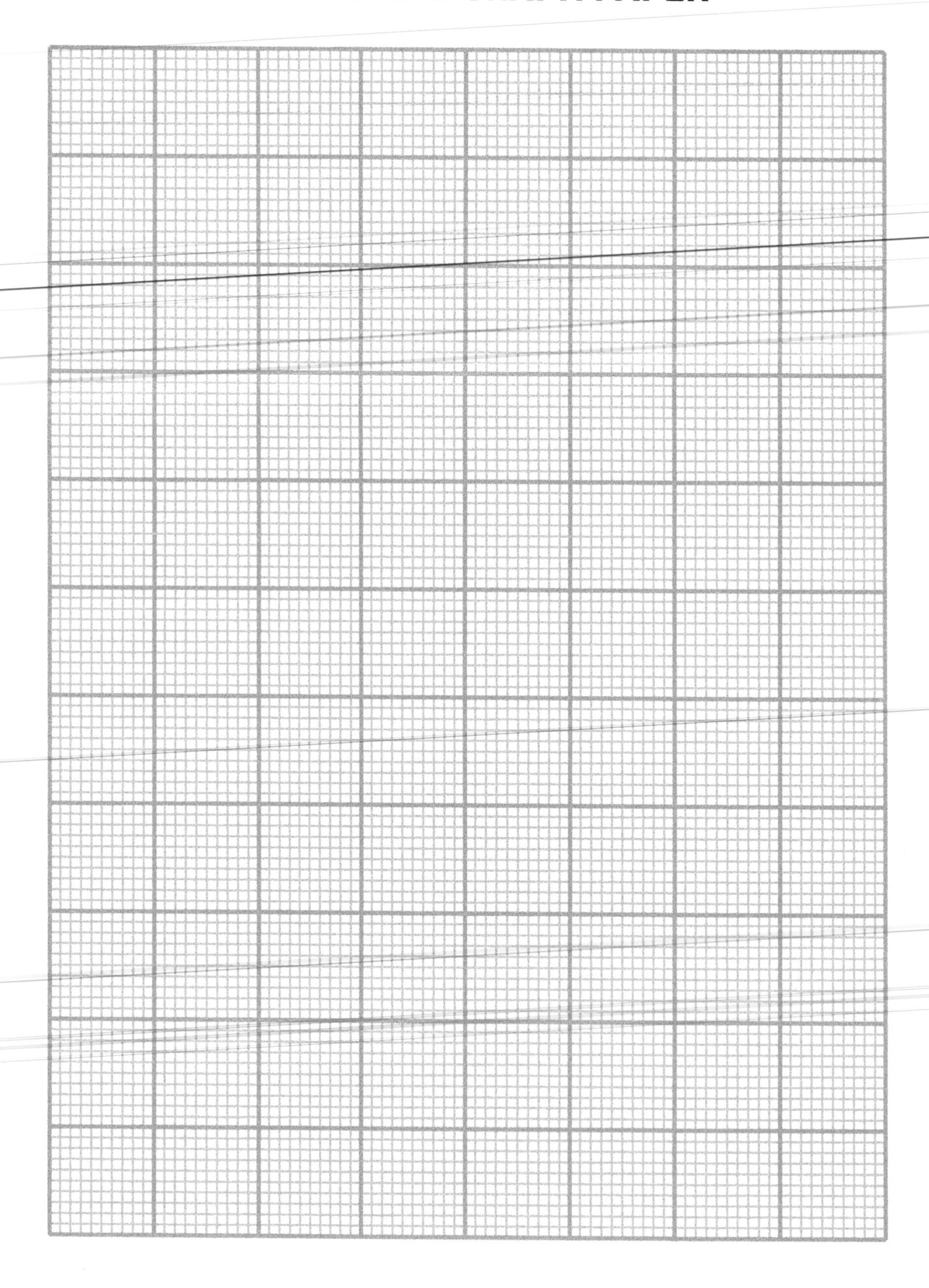

ADDITIONAL GRAPH PAPER

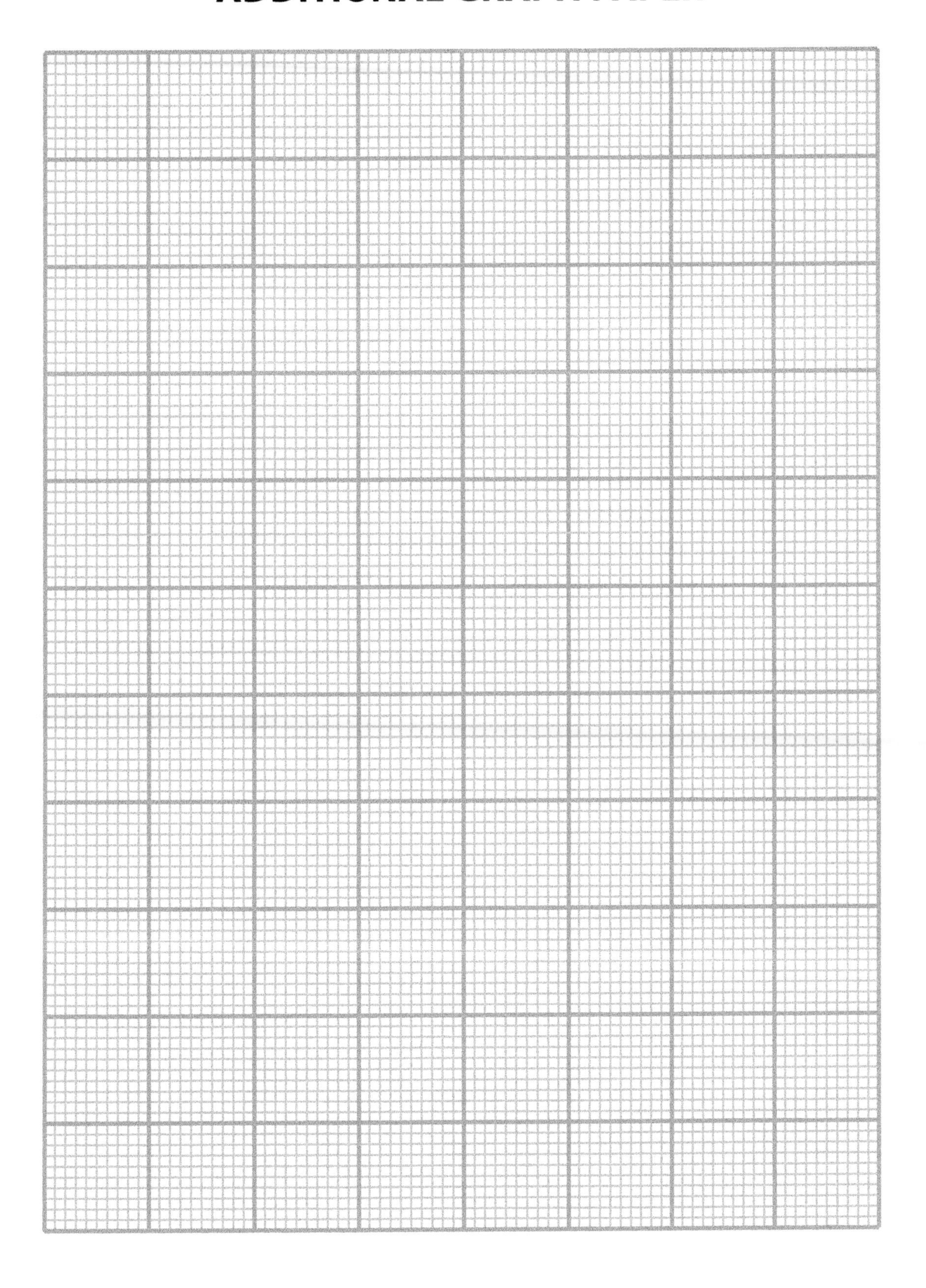